THE PACKAGING DEVELOPMENT PROCESS

THE PACKAGING DEVELOPMENT PROCESS

A Guide for Engineers and Project Managers

Kristine DeMaria
Packaging Consultant

Foreword by
Theron W. Downes, Ph.D.
School of Packaging
Michigan State University

CRC Press
Taylor & Francis Group
Boca Raton London New York

CRC Press is an imprint of the
Taylor & Francis Group, an **informa** business

CRC Press
Taylor & Francis Group
6000 Broken Sound Parkway NW, Suite 300
Boca Raton, FL 33487-2742

First issued in paperback 2019

© 2000 by Taylor & Francis Group, LLC
CRC Press is an imprint of Taylor & Francis Group, an Informa business

Library of Congress Catalog Card No. 99-67777

No claim to original U.S. Government works

ISBN-13: 978-1-56676-801-6 (hbk)
ISBN-13: 978-0-367-39919-1 (pbk)

Visit the Taylor & Francis Web site at
http://www.taylorandfrancis.com

and the CRC Press Web site at
http://www.crcpress.com

ONE OF THE measures of the maturity of a discipline is the availability and usefulness of textbooks in that discipline for undergraduate education. By that measure, packaging is still a young discipline but is maturing rapidly. It is probably true that the first texts to appear in an emerging area of study will either be rather general in nature or quite specific to those technical areas where much can be borrowed from existing disciplines. Good packaging texts exist for introductory courses and for discussion of materials, properties, and fabrication. Some texts are also available for specific areas such as food packaging and flexibles. It is more difficult and takes longer to prepare texts that will integrate all of the necessary underlying skills, technology, and knowledge for a capstone course. We have been teaching a capstone course at the School of Packaging at Michigan State University for more than 35 years. We define capstone as a project-oriented course in which students have a chance to integrate and apply knowledge gained in earlier packaging courses as well as from supporting disciplines such as physics, chemistry, business, marketing, advertising, finance, etc. We have been doing this without a textbook, and at last a useful text has been produced.

The Packaging Development Process by Kristine DeMaria provides a very useful compilation of experience, practical knowledge, and procedural guidelines to make the student's journey from academia to the real world a little easier and more productive.

One of the major difficulties in producing such a text is that there is no single agreed upon approach to the product package system development process and that further, the process to be followed will vary depending on the complexity of the project being undertaken. The author of this book does an admirable job of outlining the major steps in the package development process and then uses three separate case studies to show how and when each of the techniques required should be applied. Starting with the planning phase and continuing through initiation, concept identification, feasibility assessment, consumer testing, and final evaluation, the case studies provide a framework for students' understanding of the myriad of possibilities for the real-world projects. The three case studies which are used include (1) a crisis reaction to improperly functioning components on a packaging line, (2) a long-term productivity improvement project, and (3) a search for a new packaging concept. Each varies with regard to its complexity, the size and nature of the team that is required to address the issues involved, the financial inputs that will be required, and the length of time needed to achieve results.

In each scenario each phase of the project is discussed with the required inputs. Attention is paid to planning, producing the proper team, situation analysis, etc. Techniques from related disciplines including project management, creativity problem solving, marketing, etc. are referred to but are not presented in detail here. This is quite acceptable as a number of very good references are already in the literature. By showing which of these techniques will be needed in each given project, students will gain a great deal from the business experience of the author. The discussion of proving functionality and final package launch will be a great benefit to many instructors who perhaps have not had personal experience at that phase of the activity themselves.

THERON W. DOWNES, PH.D.
School of Packaging
Michigan State University

HAVING THE KNOWLEDGE provided by *The Packaging Development Process* will give packaging professionals the information needed to complete packaging projects with a professional approach and high degree of business savvy. Early in a packaging engineer's career, impressions and perceptions are formed that will affect future career opportunities. Being technically sound is extremely important, but equally important to one's career is having the professional know-how and an innovative approach that provides confidence and wins respect from managers and co-workers.

Packaging projects can vary greatly in complexity. Some projects will be quite simple while others will demand a lot of planning, research, testing and modifying and may take two to three years before the package is ready to be produced and marketed. *The Packaging Development Process* breaks a project into actionable steps, making an overwhelming workload manageable. Taking a logical approach to project work diminishes the probability of overlooking a task or a potential problem.

The steps for completing a packaging project are explained in the following chapters and are depicted in a flowchart on the following page. Some of the steps will not be necessary for every project. A project to modify a corrugated shipper for increased strength will use only a few of the steps, whereas introduction of an all new package may use all of the steps.

The people involved in a project and internal company policies are organization specific, but generally, certain steps are taken when working on a packaging project. For instance, the package graphics may be the packaging engineer's responsibility in one company but a separate de-

Packaging Development Process Flowchart

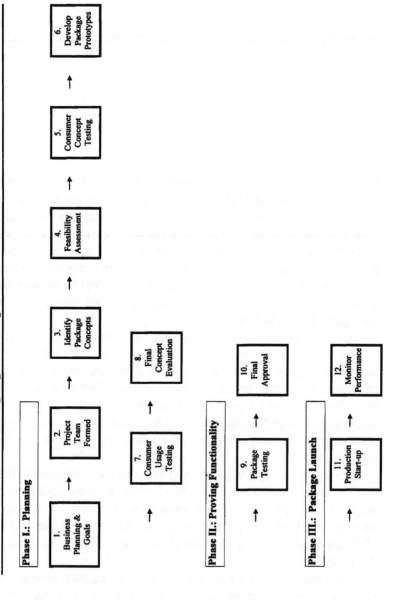

partment's in another. If a new piece of equipment is necessary, the packaging engineer may be required to source, qualify, and purchase it, or he/she may need to work with a separate engineering group that handles the details of buying all equipment.

The larger and more departmentalized a company, the more formal the methods of communication and decision making tend to be. Written communication is often the norm in a large company, whereas verbal discussions and decisions may be quite acceptable for smaller companies. Project pace also tends to be slower and risk acceptance lower in larger, more established organizations. Whatever the organizational culture, the steps and thought processes required to develop a package are similar.

CHAPTER 1

Project Initiation

BUSINESS PLANNING AND GOAL SETTING

PLANNING AND GOAL generation provide direction and vision for an organization. Business planning for many companies is an annual event that takes months to complete. Goal generation requires input from all departments, and goals are often written and rewritten before approval. The packaging department should provide input for all packaging-specific goals. Once completed, the goals are documented and distributed to those pertinent in the organization. The personality of an organization will dictate whether goals are to be strictly adhered to or if they are targets to strive for. It is wise to have a good understanding of accountability prior to submitting goals.

PROJECT TEAMS

Project teams are composed of people from the cross-functional areas needed to complete all the tasks required for a project. A team will have core members who must participate continuously for the project to thrive and secondary members who will come and go as their services are required. Sometimes packaging suppliers are viewed as secondary members on a project team.

A team needs a leader, and on packaging-focused projects, the leader is usually a packaging engineer. The packaging engineer may not be the most experienced member of the team, but it will be the engineer's responsibility to manage the team and the project. Successful teams are

1

focused with team members agreeing to do their share of project responsibilities.

Project Charter

Creating a project charter is a good planning process that enables the team to focus and think through the tasks that must be done to complete a project. In the process of writing a charter, the team provides the business need that justifies doing the project and outlines all factors relating to the project's completion. Writing a charter can be a time-consuming and difficult process, and one that many would like to skip in order to begin working on project tasks. The process is worth the effort because without thoughtful planning much time and energy will be wasted. The format for a project charter is provided in Figure 1.1.

Situation Analysis

The marketing and packaging team members often develop the situation analysis together. Marketing will contribute the financial and market purposes for the packaging project, and packaging will provide information on the technical need. A good situation analysis will help the team understand the importance of the project and prioritize it in comparison to others. Factual information will make the analysis more meaningful. If the project is quality oriented, state the extent of the problem numerically. Instead of writing, "extensive damage is occurring during shipment," determine the details of the damage. A more meaningful statement would be, "On an annual basis, 33% of all product experiences damage during shipment resulting in $1 MM worth of damaged product." A thorough situation analysis may also uncover reasons to prevent a project from proceeding, thus saving time and money.

Critical Success Factors

Critical success factors are the criteria that must be met in order for the project to meet the stated goal. These factors are determined at the initiation of a project and will not change unless the project goal changes.

(Project Name) Charter

Project Goal: Describe the purpose and objective for the project.

Situation Analysis: Describe the current technical and marketing reasons that justify the project.

Critical Success Factors: Determine which critical factors must be successfully completed in order for the project to meet the project goal.

Milestones: Provide the time and events schedule for all tasks required for project completion. Assign responsibility for all tasks.

Assumptions and Risks: List any assumptions that are being made and any risks that there may be in proceeding with the project. State any risks there are in not doing the project.

Team Members and Roles: List all team members and their roles on the team.

Team Rules: Document all agreed upon team rules such as meeting attendance, communication processes, etc.

Figure 1.1. Team charter format.

Project Timelines

Input and agreement from the entire project team is necessary on project timelines. A project timeline is a detailed schedule of all the events that must be completed for the project. There will be a series of critical events in the timeline. Without the completion of a critical event, the project cannot move forward. If a critical event is delayed, the entire project is delayed. Critical events should be highlighted to keep the team focused on them. As team leader, the packaging engineer is the keeper of the timeline. Keeping a project on schedule can be difficult. Assigning responsibility and following up with team members on the critical events can help keep the project on schedule. A good team rule is to require that project delays be reported to the team leader as soon as they are known.

When a project is new, the specific plan for the package may not yet be determined. The timeline may be somewhat vague at this point with only some major milestones listed. Once the team is focused on a specific package idea, a very detailed timeline can be established. Realistic, conservative time frames should be used, possibly even adding in a little buffer just to be safe. If the film printing process takes six to eight weeks, eight weeks should be stated on the timeline. It is better to finish on time than to be constantly explaining why the timeline is being extended. Management often feels a timeline is too long. Having solid reasons for the stated dates will help alleviate those feelings.

Project timelines can be shortened as long as the risk associated with doing so is understood and acceptable. Testing can be eliminated or one concept can be pursued rather than evaluating several, but the risk of such decisions should be well documented and communicated. Certain situations, particularly if the company is in a defensive position, may warrant taking risks in order to enter a new package into the market quickly.

The timeline is to be documented as part of the team charter, but since it will be the portion of the charter most utilized, it is a good idea to make a separate document of just the timeline for easy access. The timeline will often have to be revised as events happen and new information is learned during the course of the project. There are several formats that can be used for a timeline. The events can be shown in a table with columns listing tasks, timing, and responsibility (Table 1.1). This format is particularly convenient for projects that have a great number

Table 1.1. Timeline in Table Format.

Task	Timing	Responsibility
Idea Generation:		
team field trip	2/28	Mary
conduct brainstorm meeting	3/1	Mary
supplier brainstorm meeting	3/3	Mary/Jim
select concepts to research	3/8	Team
Feasibility Assessment:		
supplier discussion	3/9–3/24	Mary
plant trip	3/16	Mary, Doug
lab tests	3/10–5/1	Mary
quotes obtained	4/20	Mary
team meeting/ideas selected	5/1	Team
Consumer Testing:		
concept boards created	6/1	Gary
test fielded	6/4	Gary

of tasks that will take a very long time to complete. For clarity, it helps to segment the timeline into the various phases of the project.

Events can also be displayed to show the progression of the events over time (Table 1.2). This format is good for showing the cause and effect of the various events, but it may be cumbersome in size if the timeline stretches over a two- or three-year time period. Note that italic type is used for one of the tasks to indicate that it is a critical event.

Team Meetings

Meetings are effective tools for communicating and decision making. They can also be time wasters. Meetings that have purpose and are well

Table 1.2. Timeline Visually Depicting the Progression of Time.

Task	Timing					Respon-sibility
	June	July	Aug.	Sept.	Oct.	
Proving Functionality:						
Production trial	6/15					Mary
pallet ship test	⊢———⊣					Mary
physical testing	⊢—————————————⊣					Mary
consumer use testing	⊢————————⊣					Bob
final concept selected				9/18		Team

planned are effective. Discussing expectations or even specifying meeting rules in the team charter can be helpful. It may be assumed that everyone will attend meetings, show up on time, and come prepared, but verbalizing the assumptions can help make them a reality. Being flexible to everyone's schedule and accounting for a team's personality can also make for more effective meetings.

Team members may not all be located in the same building or even the same state, making meetings difficult. Telephone conferencing or videoconferencing can be a good substitute. Sending memos and project updates via the internet keeps team members informed in a timely manner. Distributing a meeting reminder as well as a copy of the agenda prior to a meeting will keep meetings organized and running smoothly. An agenda is a list of topics to be covered in a meeting. To keep the meeting within its scheduled time frame, the agenda should indicate how much time each topic is allotted and who is responsible for discussing the topic.

Project X Agenda
Feb. 6, 1999
10:00 - 11:00

scribe: Bob
timekeeper: John

Topic	Time	Who
• Lab test results	10 min	Mary
• Consumer test schedule	15 min	John
• Graphics changes	10 min	Bob
• Timeline adjustments	20 min	All
• Recap	5 min	Mary

Figure 1.2. Example of a meeting agenda.

Appointing a timekeeper to tell the group when too much time is being spent on a topic helps to keep meetings on schedule. Having a scribe also ensures that all the notes, decisions, and next steps are recorded. Allowing a few minutes at the end of the meeting to assign a timekeeper and scribe for the next meeting helps the next meeting get started quickly. It is also a good idea during those final minutes of the meeting to reiterate any decisions that were made, what tasks must be done prior to the next meeting and who is responsible for those tasks.

PROJECT INITIATION: CHAPTER SUMMARY

- Business planning is an annual event in which project goals and company vision are provided from all facets of an organization, agreed upon, and documented. The plans provide direction to the organization for the coming fiscal year.
- Project teams are composed of people from the cross-functional areas needed to complete all the tasks required for a project.
- Core team members are those whose continuous participation is required for the project to thrive.
- Secondary team members participate for only a portion of the project and thus, provide their services during specific time periods when needed.
- A project charter is a document agreed upon by all team members that provides a framework for a project stating the project goal, situation analysis, critical success factors, milestones, assumptions and risks, team members and roles, and team rules.
- A project timeline is a detailed schedule of all the events that must be completed for a project.
- Critical success factors are the criteria that must be met in order for the project to meet the stated goal. These factors are determined at the project's initiation and will not change unless the project goal changes.
- A critical event is one in which without its completion the project cannot move forward. If a critical event is delayed, the entire project is delayed.
- An agenda is a list of topics to be covered at a meeting.

PACKAGING PROJECT EXAMPLES

Scenario A: Credo's Productivity Project

Bill is a packaging engineer at a food company and is responsible for the Credo's product line. His manager stops in his office one Monday morning to tell him their division has been requested to provide productivity savings. She requests that Bill explore savings options on the film used on Credo's. Bill and his manager briefly discuss the project, then she asks Bill to pull together a project team and get back to her next week with a plan. Bill phones his marketing counterpart, John, after his manager leaves. John has also just been made aware of the project. They schedule a meeting to begin the charter process. To make the meeting more productive, they each agree to rough out a charter prior to meeting. Bill hangs up and gets to work.

Bill knows this will be a very technical project, and although he will work with a team of people to accomplish it, he will be responsible for the bulk of the work. Based on his knowledge of the current package and product requirements, Bill roughs out a charter. He and John meet and combine their efforts. They also determine who should be involved on the project team and call a meeting. The purpose of the meeting is to present the project and to discuss the charter for team consensus. Bill coordinates a meeting time for all the team members and sends out an agenda. He attaches a copy of the charter and requests that everyone look it over and come prepared to discuss it. Once the team agrees upon the charter, the first step of the project will have been completed. The charter will keep the team focused on one goal, and each member can contribute his/her part. The team can determine a broad timeline based on past experience and certain assumptions. Their timeline will become more detailed once the specific film productivity options have been determined. The timeline may change again if difficulties are encountered as the project progresses. The Credo's project charter is provided.

Credo's Film Productivity Charter

Project Goal

To identify and capture Credo's packaging film productivity savings with a target of $2.8 MM annually.

Situation Analysis

Packaging productivity savings are being sought in order to maintain Credo's retail price. Manufacturing costs are estimated to increase 25% ($2.8 MM) more in the next two years than they have in the past due to necessary changes in the production process for the product. The increased cost will have to be passed on to the consumer unless savings can be realized. Credo's is a very price-sensitive product. Consumer research indicates that 30% of Credo's consumers base their purchase intent on price alone. A Credo's price increase is expected to result in a loss of consumers to the competition.

Critical Success Factors (CSF)

A film change must

- provide $500 M or more in savings to warrant dedicating resources to the project
- protect the product at parity or better to the current package
- not affect the overall aesthetics of the package graphics
- not affect the product's quality perception with consumers
- be completed prior to the manufacturing cost increase (18 months)

Milestones

Task	Timing	Responsibility
Film supplier meetings	2/1–2/18	Bill
Brainstorm meeting	2/4	Bill
Ideas determined	3/1	Bill/Team
Feasibility assessment	3/1–5/10	Bill
Order and produce test film	5/10–7/10	Bill/Kerri
Plant trial	7/14	Bill
Physical testing	7/14–11/14	Bill
Shelf life/sensory testing	7/14–12/14	Bill/Ted
Consumer testing	7/30–8/30	Marge
Data evaluation/ Go decision	12/15	Team
Extensive line trials	12/20–1/20	Bill/Sandy
Team Go or Stop decision	2/1	Team
Conversion complete	4/1	Kerri

Assumptions and Risks

- Current packaging manufacturing equipment is to be utilized. No funds are available for alternate equipment, and timing does not warrant it as an option.
- The provided timeline does not account for any potential modifications to the test films and subsequent required testing.
- If savings cannot be realized, 30% of the consumer base could be lost.

Team Members and Roles

Bill Smith	Packaging
John Tames	Marketing
Sandy Lane	Operations
Kerri Alt	Purchasing
Ted Junid	Product Development
Marge Landers	Consumer Behavior

Team Rules

As team leader, Bill agrees to relay project information via a monthly update report and schedule meetings when critical events in the project make them necessary. The team agrees to communicate any timeline alterations to the team leader as soon as they are known.

Scenario B: Oula Package Redesign

Susan is a packaging engineer responsible for the Oula toiletry product line. One of the Oula products is a dual moisturizing cream. The current package is a divided container with a flip-top closure. The package is one piece formed out of plastic. To use the product, the consumer opens the flip-top to expose two wells, each filled with a different moisturizing cream. The consumer dips a finger into a well to remove cream and applies it to the skin. The creams are applied first one then the other to the skin and must be used in conjunction for maximum effectiveness.

Consumers have lodged several complaints regarding the packaging. First, they do not feel the flip-top closure is secure enough and do not trust it to stay closed in their handbag or overnight bag. The two-well container is an issue because eventually the cream from one well gets into the other well, and consumers don't always finish the two creams at the same time. The consumers did state that they like the fact that the two creams are always together since that is how they must be used.

The Oula product line is viewed as one of the company's rising stars. Sales for the moisturizing cream are increasing, but the percentage of consumer complaints related to packaging are increasing at a faster rate. Many consumers say they would switch products because of the packaging if they could find another moisturizing cream that works as well. Susan feels that issues with the packaging will escalate as their consumer base grows. She is in the midst of writing project goals for next year and would like to include a new project directed at improving the Oula moisturizing cream package. Susan discusses her idea with Chris in marketing who agrees with Susan's thoughts. The Oula product line manager is also highly in favor of exploring new packaging concepts for Oula moisturizing cream.

Susan and Chris phone their cross-functional counterparts and schedule a meeting with the purpose of writing a project charter. After a few meetings and much discussion, the team agrees upon the following charter.

Oula Dual Moisturizing Cream Packaging Redesign Charter

Project Goal

To redesign the Oula dual moisturizing cream packaging so that closure integrity and product containment become complementary features to the consumer.

Situation Analysis

Consumer complaints regarding packaging jumped 20% last year as sales volume increased 10%. Sales volume is anticipated to increase 8% in the next year, and subsequently, packaging related complaints are anticipated to increase 20–25%. The company goal of decreasing consumer complaints and the threat to Oula's sales growth warrant a thorough exploration of packaging concepts that better meet the needs of the consumer.

Critical Success Factors (CSF)

- The redesigned package must be rated by consumers as a significant improvement versus the current package for closure integrity and product containment.
- The redesigned package must be rated by consumers as an improvement versus the current package for overall liking of the packaging.
- The redesigned package must protect the product at parity to the current package.
- Management agrees to approve up to a 10% packaging material cost increase given the redesign meets all CSF.
- Additional cost for tooling or equipment may be acceptable given the redesigned package meets all CSF.
- Management approval for a redesign is needed by 6/1 in order to obtain any funding from this year's capital budget.

Milestones

Task	Timing	Responsibility
Idea generation	1/3–2/10	Susan
Ideas determined	2/12	Susan/Team
Feasibility assessment	2/12–4/1	Susan
Concepts selected	4/10	Team
Concept boards created	3/20–4/10	Susan/Barb
Consumer concept testing	4/14–5/10	Barb
Management approval meeting	5/15	Team

Assumptions and Risks

- Although sales continue to increase, the team feels that the current package will eventually have a negative effect on sales unless it is redesigned.
- If sales increase as anticipated, an additional packaging line will be required. The team feels that any capital fund investment for Oula should be applied toward improving the product line.
- If needed, a co-manufacturing location will be allowed for production of a new package.

Team Members and Roles

Susan Shelton	Packaging
Chris McCane	Marketing
Paul Willis	Operations
Mark Hilk	Purchasing
Barb Hawkins	Consumer Behavior

Team Rules

The team will make the packaging redesign project their number one priority. All team members will participate, if needed, in the creation of package concepts for consumer testing.

Scenario C: Carton Crisis

Paul from the manufacturing facility phones Kathy, the packaging engineer who works on his product line, and explains to her that they have been having issues with a carton. The cartons are continuously jamming in the machine, causing a great deal of downtime. After asking more questions, Kathy learns that the problem began when cartons from lot KL812 were placed in the machine. The line had been running with no issues when using cartons from lot KL510. It appears that two and three cartons at a time are feeding into the mechanism that sets up and glues the individual cartons. The technical crew at the manufacturing facility has tried to adjust the machine but cannot prevent the problem from occurring. Paul indicates that he cannot see a difference in cartons from the two lots and has even measured them, but the dimensions are the same.

Kathy requests manufacture dates for the cartons and is told that KL812 cartons were produced in August, and KL510 cartons were produced in July. Kathy suggests they try cartons from another lot if any are available, while she calls the carton supplier to get more information. She also asks Paul to express mail cartons from each lot, KL510 and KL812, to her.

A project charter and a project team are not needed for this situation.

Kathy wants to quickly inform her manager and her operations and purchasing counterparts. So, she leaves them a group voice mail message explaining the situation and what steps she is taking indicating that she will call them individually when she has talked to the carton supplier. She then proceeds to phone the carton supplier.

CHAPTER 2

Identify Package Concepts

PACKAGING PROJECTS ARE born in a number of places inside and outside of the work organization. People deal with packaging in some aspect most every day of their lives, from the cereal box used in the morning to the corrugated shipper that transports their mail order item to their mailbox. This familiarity with packaging tends to broaden the field of input when packaging opportunities are being discussed.

Companies often categorize projects based upon the project's goal. A packaging engineer might work on projects in all areas and may even be required to do so. Project category names may vary depending on the organization.

PROJECT CATEGORIES

Innovation and Growth

These packaging projects utilize new marketing and technical ideas to reposition a stagnant product or deliver on unmet consumer needs, all with the ultimate goal of increasing product sales or establishing a product as a leader in its field. They are often the most challenging and organizationally visible projects.

Productivity

Delivering savings on an annual basis is often a required goal in many organizations. Unfortunately, packaging is not exempt from these goals. Savings may come from a change in packaging material or a component change or a change in the packaging manufacturing process. The end result may even be invisible to consumers but can take just as much time and effort to qualify.

Quality

Packaging quality projects may involve improving production efficiency, reducing product damage, correcting a packaging defect, improving a packaging feature, or extending product shelf life via packaging barrier improvements. Hopefully, all quality projects are planned in an effort to meet the standards set by the company. It is extremely costly and exhausting to correct a defect on a package that has been released to the market.

Defensive

Some projects result when a company's competition releases a new package or product into the marketplace that could endanger its product's market share. The threat is assessed and, if necessary, resources are rallied to quickly respond.

Brand Maintenance

These are smaller product line support projects or unplanned projects that crop up throughout the year. An example would be enlarging a shipper width 1/4″ to accommodate an equipment change made at the manufacturing facility. Forming a project team may not be necessary for many brand maintenance projects.

WHERE IDEAS COME FROM

Innovative ideas are more likely to occur through active searching. Some ideas will be generated through the natural course of doing business, but as the corporate world becomes more and more competitive, it is even more important to actively seek out ideas. The packaging engineer is familiar with his/her company's manufacturing process, emerging packaging technology, and most often has packaging peers to spur on ideas. But, packaging ideas can come from many different sources inside and outside of a company.

Packaging Suppliers

Utilize packaging suppliers as an external source for ideas by requesting that they generate three new ideas of their own for a specific project. Confidentiality agreements between a company and its packaging suppliers allow them to share confidential information regarding their processes and project goals. Touring a supplier's facility and gaining knowledge of their processes can also generate ideas. Shared information and open communication leads to a greater understanding of each other's needs and thus, to more ideas and problem resolution.

Many companies have exclusive purchasing contracts, sometimes called strategic alliances, with specific packaging suppliers. Strategic alliances specify material cost and are usually time specific with terms for ending the relationship if project goals cannot be met. Although strategic alliances are formed with only a few suppliers, relationships with competitive suppliers should be established and maintained to keep up to date with evolving technology. A narrow supplier focus can limit the pool for ideas.

Consumer Testing

Going directly to the source via consumer testing will provide information on how consumers use and feel about a package and product. Consumers are usually quite agreeable to providing their thoughts on improving or changing a package. Asking consumers to provide information about competitive products can also generate new packaging ideas.

Generating Ideas through Brainstorming

Conducting a well-planned brainstorming session is an active method for generating a lot of ideas. The meeting will be successful if the leader, usually a packaging engineer for packaging projects, has taken the time to prepare several brainstorm activities and has invited a cross section of people within the company to attend. Anyone can contribute. Inviting people who are not directly involved with the package or product being discussed can bring a different perspective. If appropriate, packaging suppliers can also contribute by attending.

Establishing a fun atmosphere for the meeting will help the creative thoughts flow. Snacks and beverages and music can add to the relaxed setting. Starting the meeting with an icebreaking event can relax the group and acquaint its members. An icebreaker is an interactive group activity that aids in making group members feel more friendly and like a cohesive unit. An example of an icebreaker would be to ask everyone to secretly write down something unknown about themselves, such as, "my great grandfather spent a year in a circus" and "I love to garden," and give it to the meeting leader. The meeting leader would then read off these items and everyone must try to match the statements to the person who wrote them. Icebreakers that involve team activities such as a competition to see who can build the tallest tower using gumdrops and dried spaghetti noodles are fun and immediately engage everyone in the meeting. If the brainstorm meeting needs a momentary topic change or a refurbishment of energy, icebreaker activities can also be conducted during the meeting.

After the icebreaker, the topic that is to be brainstormed is explained by the team leader. The first portion of the meeting is for idea generation. In the second portion, the ideas are evaluated to determine which ones warrant further investigation. The goal during the idea generation portion is to generate as many ideas as possible and to do so without judgment. Meeting participants should build off of each other's ideas. All ideas should be encouraged and treated as valid. There are no bad or stupid ideas. Flip charts help to get ideas written down quickly where everyone can see them. As the pages fill up, they can be hung on the wall. Samples of packages or containers relating to the discussion topic are useful props for thought generation.

The topic can be discussed directly using open-ended questions, but a more effective method is to use creative thinking activities designed to spur innovative thought. Some of the exercises may seem foolish or off the wall, but if one or two good ideas can be identified, the exercises are well worth the effort.

Brain Purge

Starting with an activity called brain purge, in which participants write ideas on paper without talking, is most effective in producing a large number of ideas. For this activity, each person is provided with a piece of paper and is given two minutes to write down ideas. The papers are then passed to the person to the right for another two minutes and so on. This way participants can build off of each other's ideas. When everyone has exhausted their ideas, verbal discussion for further brainstorming and clarification is done. Once the brain purge activity is completed, any of the following activities can be done in any order.

Assumption Reversal

Assumption reversal is a technique created by Steve Grossman (1984) in which all assumptions related to the topic are listed no matter how obvious they are. The assumptions are then reversed to stimulate new ideas.

For example, a group is generating packaging ideas for canned cat food.

Assumptions	Reversals
Cats eat cat food.	Cats don't eat cat food.
Cats eat food from a dish.	Cats don't eat from a dish.
Cats can't open the can of food.	Cats open their own food.
Cat food doesn't smell good to humans.	Cat food smells good to humans.
Cats tend not to eat leftover cat food.	Cats eat all of their food.
The package is tough and sturdy.	The package is soft.
The package requires a can opener.	The package opens without a tool.
The package contains and protects the cat food.	The package doesn't hold and protect the cat food.

Ideas that are generated include the following:

- provide recipes so pet owners can make their own cat food
- market a nutritional powder that owners use when making their own cat food or cat treats

- design a package that cats eat from directly, which can then be thrown away, eliminating the need to clean a messy cat dish
- design a package with a timer that dispenses food for a week at a time
- provide cat food in single-serve packages sized for the small or large cat appetite
- use a plastic cup or a pouch instead of a can
- provide an easy open feature on the package
- create an edible package
- design a package that entices cats to play so they get some fun and exercise while opening the package to eat

Semantic Intuition

Another technique, which originated at the Battelle Institute in Frankfurt, Germany (Warfield, Geschka, & Hamilton, 1975), is semantic intuition. For this technique, the process used to name products is reversed. Instead of naming a product after it is invented, it is first named and then invented. To do this activity, generate two sets of words related to the brainstorm topic. Then randomly combine one word from one set with a word from the other set. The word combinations are used to stimulate new ideas.

For example, a group is brainstorming for new packaging and product ideas for chewing gum.

Uses for Gum	Parts of Gum
chewing	wrapper
snaps	stick
freshens	overwrap

Word Pairs	Ideas
freshens-wrapper	flavored wrappers keep gum highly flavored
chewing-wrapper	edible wrappers
snaps-stick	flavor crystals burst and snap when chewed, or one large stick of gum scored so that portions can be snapped off
snaps-overwrap	closure snaps over sticks of gum

Product Improvement Checklist (PICL)

Product improvement checklist was developed by VanGundy (1985) in which a poster containing 576 stimulus words organized into four categories is used to generate ideas. The four categories are

(1) Try to . . . (e.g., sketch it, wipe it, tighten it)
(2) Make it . . . (e.g., soft, transparent, magnetic)
(3) Think of . . . (e.g., escalators, oatmeal, time bombs)
(4) Take Away or Add . . . (e.g., layers, sex appeal, friction)

One word is randomly selected from a category to see if it develops into a new idea. Combining words within or between categories can also be done. The poster is not necessary for this activity. Words can be selected randomly from the dictionary instead.

To illustrate, suppose the brainstorm topic is to develop ideas to portion control a powdered drink mix.

> *Make it . . . a stick* becomes an idea to convert the powder to a solid on a stick like a popsicle that would be stirred in a quart of water until it dissolved to make the perfect refreshing drink.

> *Try to . . . explode it* becomes an idea for shaping the powder drink into disks (like those used for upset stomachs or denture cleaner) that fizz and dissolve when dropped into a glass of water.

Association Technique

Association technique is a method in which a topic unrelated to the meeting topic, usually a well-known person, place, movie, or thing, is discussed and then related back to the meeting topic. A topic is chosen, then the brainstorm participants make a list of items related to that topic. Each of those items is then associated to the brainstorm topic.

For example, the brainstorm discussion is centered on taking a shampoo product and making it appealing to children via packaging. The group chooses New York City and makes a list of items associated with it. They then use the list and try to make associations back to the shampoo for children.

New York City	Children's Shampoo Package Association
Musicals	When the bottle is squeezed or cap is opened, the package plays a tune.
Tall buildings	Shape the package like a building with a clear strip in the container that is the "elevator." Kids watch the elevator move down floors as the shampoo empties. When it hits the first floor, it is time to buy more shampoo.
The Village	Packages would be shaped like various buildings in a village—kids can collect them all.
Wall Street	Utilize the package in a game to win prizes.
Subway	Shape the package like a train.
Madison Avenue/shopping	Incorporate a handle like a shopping bag in the package for easy use.
Food/restaurants	Use postconsumer packaging to make the shampoo containers; get kids involved in a recycling program.
Dance clubs	Have flashing, multi-colored jewels on the package.
Taxicabs	Shape the package like a car.

Do this several times with various topics. Many of the associations may seem foolish, but the exercise will definitely stretch the participants' minds into different thought directions.

Once all the ideas are exhausted, the second portion of the brainstorm meeting is spent evaluating the ideas to determine which ones will be researched further. Now is the time to voice opinions. The ideas can be rated by going through them one at a time or having everyone place an "x" by their five favorite concepts on the pages hanging on the walls. The group can discuss the "winning" ideas more and then agree if they are to be further investigated for feasibility. In some instances, the evaluation process goes more smoothly if the ideas are categorized into several groups before the top ideas are selected.

Specific Problem Resolution

Sometimes a very specific packaging problem may arise in which technical information and expertise are needed to brainstorm potential causes. For instance, a pressure sensitive label no longer adheres when exposed to refrigeration temperatures. The packaging engineer determines potential causes through a very logical brainstorming process and then uses that information to resolve the problem. The logical process involves focusing on the materials and processes involved in the manufacturing of the packaging item and those used in package production. Creative thinking techniques designed to produce novel concepts are less useful in these situations.

Field Trips

Another method for idea generation is to take a field trip to explore packages already on the shelf. A trip to a store that stocks items unrelated to the package being studied can expand the idea possibilities. If a new package concept is being explored for a food product, a package found at a hardware store or a cosmetic counter may provide new ideas. Understanding what the competition is using for packaging can also be very useful information.

A field trip can be taken prior to a brainstorm session to collect packaging props for the meeting. Unusual and unique packages in addition to packages directly related to the discussion topic will help generate ideas. It may also help to ask those attending the brainstorm meeting to take a field trip of their own. If time allows, a field trip can be taken as part of the brainstorm meeting.

A Final Note on Generating Packaging Ideas

Use of these methods will generate packaging ideas and new concepts. These methods should not be placed on a shelf and pulled out when a specific project has been identified but they should be seen as tools for continuous use. Actively seeking creative ideas can take a lot of energy, but being an innovative person is a vital asset to any company. Scheduling creative thinking activities into the weekly work plan or making them part of annual performance goals can ensure their use.

IDENTIFY PACKAGE CONCEPTS: CHAPTER SUMMARY

- Anyone within a company or anyone who uses a product or package can generate ideas for changes or improvements.
- Projects can be categorized based on the project goal as innovation and growth, productivity, quality, defensive and brand maintenance.
- Confidentiality agreements are legal documents agreed upon between two parties stating that all said and seen will be kept confidential or secret and will be discussed only between those signing the agreement.
- Strategic alliances are exclusive purchasing contracts that specify packaging material costs and are usually time specific with terms for ending the contract if project goals cannot be met.
- Packaging suppliers, consumers, brainstorm meetings, creative thinking activities, and field trips can help generate innovative ideas for packaging programs.
- An icebreaker is an interactive group activity that aids in making group members feel more friendly and like a cohesive unit.
- The brainstorm creative thinking activities are as follows:
  (1) *Brain Purge:* Participants write ideas on paper without talking for two minutes, then pass their paper to the person on their right for another two minutes, and so on.
  (2) *Assumption Reversal:* All assumptions related to a topic are listed and then reversed to stimulate ideas.
  (3) *Semantic Intuition:* For this technique, the process for naming products is reversed so that the product is named and then invented.
  (4) *Product Improvement checklist (PICL):* Four categories, Try to . . ., Make it . . ., Think of . . ., Take Away or Add . . ., are combined with randomly selected words to generate ideas.
  (5) *Association Techniques:* A topic unrelated to the brainstorm topic, usually a well-known person, place, movie, or thing, is discussed and then related back to the brainstorm topic.

PACKAGING PROJECT EXAMPLES

Scenario A: Credo's Productivity Project

Bill schedules a brainstorm meeting for his productivity project. Because his project is quite technical, he narrows his list of meeting attendees. Bill invites those who can input on a packaging technical level—his packaging peers and the technical support group from the current film supplier. He also includes team members: John from marketing, Kerri from purchasing, and several people from manufacturing.

Because of the specific technical topic, Bill decides not to use creative thinking techniques but does bring some current package samples. He requests that the film supplier bring any applicable samples. Bill starts the meeting with introductions and an icebreaker. After explaining the project goal and criteria for success, he uses a flip chart to show a schematic of the multilayer film used to package Credo's.

polypropylene
ink
adhesive
polypropylene
heat seal layer

The group discusses the purpose for the current film configuration and delves into other possibilities. All ideas are considered and recorded in the first portion of the meeting. The list is then evaluated more closely in the second portion of the meeting.

By the end of the meeting, the group has compiled the following list and placed an asterisk (*) next to items they feel have the most probability for success.

* • Explore different polypropylene suppliers.
* • Explore alternate heat seal suppliers.
* • Decrease the gauge of one or both polypropylene layers.
* • Produce the film as a monolayer versus a multilayer.
* • Determine if there are any advantages of surface print versus reverse print.
* • Narrow the web width.

- Produce larger rolls of film.
- Use cold seal versus heat seal.
- Purchase quantities and inventory opportunities.
- Decrease the graphics ink coverage.
- Determine if there are opportunities for savings in the film making process.
*• Explore alternate materials versus polypropylene.
- Produce the film in-line at the manufacturing plant.
- Print the film in-line at the manufacturing plant.
- Contact the current suppliers of polypropylene for their input on cost savings.

There is one other thing that Bill can do to generate ideas for his project—contact other film suppliers and get their input. Bill decides to wait until the feasibility assessment is done before talking to other suppliers. He has a good relationship with his current film supplier, but if the potential for significant productivity savings does not exist, Bill will need to discuss his project with other film suppliers.

Scenario B: Oula Package Redesign

Susan schedules a brainstorm meeting. She invites all the members of her project team, several of her packaging peers, and four people from the manufacturing facility. She decides not to invite any packaging suppliers at this time because specific packaging materials have not been selected.

Susan asks everyone to bring package samples that contain two products in separate compartments to the meeting. Prior to the meeting date, Susan takes several trips to a variety of stores to obtain packaging samples. She decides to bring her portable stereo to the meeting to play inspirational music. She buys a nice selection of candy for the meeting from a candy store known for their fun and wacky products and schedules for other food and beverages to be available. She plans to start the meeting off with an icebreaker and to use several creative thinking activities to generate ideas.

After the icebreaker, Susan passes a sheet of paper to everyone at the meeting. At the top of each sheet she has written "Brain Purge," and in the upper, right-hand corner a list of the needed package improvements, closure integrity and product containment, have been written. Susan encourages everyone to draw pictures and to make whatever notes are needed to explain their package ideas. After completing several more creative thinking activities, Susan has a lot of ideas. The remainder of the meeting is spent discussing the ideas and selecting the ones that everyone feels best meet the project goal.

Scenario C: Carton Crisis

Kathy needs more information before she can proceed with an action plan regarding the issues with the cartons. She phones Sara in purchasing and together they conference call Bob at the carton supplier and explain the situation, stressing that the plant is experiencing a lot of production downtime. Bob says he will check their manufacturing records to ensure no process changes were made when cartons from lot KL812 were produced. He suggests that high humidity could be causing the cartons to stick together. Kathy doesn't feel humidity is the cause since cartons from lot KL510 had no issues and were used just prior to KL812 cartons.

Bob offers to dispatch a technical representative to the manufacturing facility immediately and says he will get carton production information and call Kathy back as soon as he can. Kathy hangs up, leaving Bob and Sara discussing the costs the manufacturing facility is incurring due to the carton situation.

Using brainstorm creative thinking techniques will not likely help Kathy resolve her carton situation. She needs technical expertise for her specific problem. She decides to discuss the situation and brainstorm possible causes with some of her packaging peers. She finds her manager in her office. They call in Ted, a fellow packaging engineer, who handles cartons for all of his products.

Ted asks to see the carton specification. He notes that the specification is quite detailed, including information such as paperboard basis weight, caliper, stiffness, and a dimensioned drawing. There is also material information about the type of board used and its coating. Shipping and storage requirements are also detailed. He asks Kathy if she has ever determined the coefficient of friction for her carton. She has not.

Together they brainstorm for potential causes for cartons jamming in the equipment. They concentrate on the equipment first. Paul from the manufacturing facility has indicated that adjusting the machine could not resolve the problem. Cartons from a different lot ran well on the equipment just prior to the problem occurring. The carton packing process has not changed recently and it did not change when cartons from lot KL812 were placed in the machine. The line operator has been running the equipment for years. It is not likely the equipment is the issue.

The three then focus on the carton for potential causes. The carton specification has not been modified for two years and the carton has never had any production issues. Kathy says she does not think the problem is related to board caliper. If the cartons were produced on a thicker paperboard than specified, the manufacturing facility could have dealt with it through adjusting the equipment. Line speed may decrease, but the cartons would not be feeding in two and three at a time and jamming. They all agree that the problem is probably not weather related because storage and handling was identical for the problem cartons and the cartons that ran well earlier in the day. They then theorize that something may have changed in the coating used on the cartons that is altering the coefficient of friction. The coating itself may be chemically different, or the coating application process may have been done differently.

Kathy's next steps are to measure the coefficient of friction on the carton samples she will receive the next day and have the coating on the two lots of cartons chemically analyzed. She also needs the carton manufacturing data from her carton supplier.

CHAPTER 3

Feasibility Assessment

STEPS TO DETERMINE IDEA FEASIBILITY

ONCE PACKAGING IDEAS have been generated, they should all be recorded in one document. List the ideas, their probability of success (POS: high, medium, or low), and any comments about the idea that were discussed during the latter portion of the idea generation session. The most probable ideas should begin the list. Distribute the list to everyone on the team and also to anyone involved with the package or product line to spur additional ideas or generate feedback regarding a particular concern.

Further assessment of the top ideas is now done to ensure they are worthy of pursuit. Use the following steps to determine feasibility:

(1) *Step 1: Determine if the idea meets the criteria for success as was determined in the team charter.*
Most often this question cannot be fully answered without committing more time and resources to explore the idea, but the probability of the idea being successful can be assessed.

(2) *Step 2: Determine the probability for success.*
Utilize all information resources and conduct necessary tests to gather specific data in order to complete Step 1. During the idea evaluation portion of the brainstorm meeting, the top ideas are highlighted and their probability for success, high, medium or low, is determined based on the team members' expertise and experience. Those assess-

ments must now be confirmed with data via information search and testing.

(3) *Step 3: Reevaluate ideas.*

Based on the new information that has been obtained, determine if the top ideas should be investigated further. Take a second look at the other ideas on the brainstorm list and be sure the decision not to pursue those ideas was correct.

(4) *Step 4: Determine if more idea generation is needed.*

If none of the ideas has a high probability for success after gathering more detailed information, more idea generation is needed.

(5) *Step 5: Determine what next steps are needed to fully qualify (i.e., proving functionality) the selected idea or ideas.*

The process of creating a timeline will provide a clear understanding of how complicated the qualification process will be and whether or not a desired completion date can be met, and will expose issues previously not considered. A timeline will also be needed at a project approval meeting.

Gathering Information

The amount of information needed may seem overwhelming. Many questions will arise when completing the steps to determine feasibility. It is helpful to ask the following questions: *"What information do I need to know to answer this question?" "Who has information that can help me?" "Where are other resources located that may have more information?"*

Explore each idea using the steps to determine feasibility, and make a list of the people to contact and the questions to ask. Note other resources such as books, periodicals, internal reports, etc., that may provide more data. The information search will lead to more questions and additional leads. Continue to ask "why" during the information gathering. Be open to all options and try not to kill an idea before it has time to grow. Perhaps one idea meets the success criteria except for manufacturing capabilities, and there are no funds for new equipment. Before eliminating the idea, discuss it further with the manufacturing location for their input, and consider other manufacturing options such as co-manufacturing or renting equipment. If an idea is too costly, explore options to cut costs before concluding that the idea is not feasible.

Some initial lab testing may be needed to gain insight on how well the package will perform. Actual packages may not be available to test, but there are often some preliminary tests that can be done to provide data indicating whether to proceed or not. Traveling to supplier locations and to manufacturing facilities may also provide the needed information. Small-scale manufacturing trials at an equipment supplier or at the manufacturing facility can often be conducted to assess feasibility. For example, using laser technology to create a pouch seal may be an idea being considered. Companies with the laser technology can simulate the process on their lab equipment using the pouch material and duplicating manufacturing procedures such as line speed as much as possible.

The most important thing a packaging engineer can do during feasibility assessment is to challenge all thought processes—theirs, coworkers and packaging suppliers. Continue to ask questions until all avenues have been exhausted.

Deciding the Project's Direction

The goal during feasibility assessment is not to obtain complete information, but to gather applicable data to justify proceeding. A lab test may prove an idea to be feasible but testing at the manufacturing facility will be the only way to truly qualify the idea. Team consensus and approval is needed before moving to the proving functionality phase of the project. Thorough information gathering will ensure that all data has been collected so the team can make educated decisions.

On some projects, consumer feedback is needed prior to selecting the final one or two ideas that will be pursued. (Consumer testing will be covered in the next chapter.) Weighing consumer preference versus project cost and timing can be difficult. A definite answer regarding a new package's ability to grow sales volume is not easily attained. Some decisions will come down to instincts. Thus, it is critical to get management's input as well as the team's.

Once all the questions are answered and the feasibility assessment work is complete, the information is to be presented to the team. The team can then select the package concept or concepts that will be pursued to accomplish the project goal. A very detailed project timeline can be created now that package concepts have been selected.

Project Reviews

During the course of a project, there are times when a project review should be conducted. At a project review, all the project facts to date and thought processes behind a project are presented to a group of people for their critique. Individuals from all divisions and levels within the company can be invited to participate in a project review. The purpose of the review is to make sure no detail has been missed. The review is an opportunity to gain from the expertise and experience of others and to get project buy-in from all divisions of the company. A project review can be very nerve-wracking, but the meetings are not to be viewed as a personal affront to a packaging engineer's abilities. The review is done to protect the company from making poor business decisions.

It is a good idea to conduct a review at critical points in the project, such as when significant resources (time and/or money) must be committed to a project, prior to exposing the project to consumers and possibly the competition, or prior to final project approval. Depending upon the nature of a project, two or three project reviews may be conducted before its completion.

FEASIBILITY ASSESSMENT: CHAPTER SUMMARY

- The five steps to determine project feasibility are
  - (1) *Step 1: Determine if the idea meets the criteria for success as determined in the team charter.*
  - (2) *Step 2: Determine the probability for success.*
  - (3) *Step 3: Reevaluate ideas.*
  - (4) *Step 4: Determine if more idea generation is needed.*
  - (5) *Step 5: Determine what next steps are needed to fully qualify (i.e., proving functionality) the selected idea or ideas.*
- To answer the questions that arise during information gathering, ask the following questions: *"What information do I need to know to*

*answer this question?" "Who has information that can help me?"
"Where are other resources located that may have more
information?"*

- The goal during feasibility assessment is not to obtain complete information but to gather the applicable data to justify proceeding.
- At a project review, all the project facts to date and thought processes behind a project are presented to a group of people for their critique.

PACKAGING PROJECT EXAMPLES

Scenario A: Credo's Productivity Project

Bill takes the list of ideas generated from the brainstorm meeting and compiles them into a memo that he will distribute to his project team. Bill uses the notes he took during the discussion portion of the brainstorm meeting to create Table 3.1 (POS = probability of success: H = high, M = medium, L = low; for timing: S = short, M = medium and L = long).

Bill now takes the top ideas and completes the five steps to determine feasibility for each idea.

Idea: Decrease Gauge of One or Both Layers of Polypropylene

Step 1: Determine if the Idea Meets the Criteria for Success

Critical Success Factors (from the team charter): a film change must

- result in productivity savings significant enough to affect the price point of Credo's
  It makes sense that using less film will result in a savings, but how much?
- protect the product at parity or better to the current package
  Will a thinner gauge film perform as well as the current film?
- not affect the overall aesthetics of the package graphics

Table 3.1. Credo's Film Productivity Project: Brainstorm Ideas.

Idea (top ideas in italics)	POS	Timing	Comments	Next Steps
Decrease gauge of one or both layers of polypropylene	H	S	Needs testing to determine film's tolerance to manufacturing, distribution, and consumer use	Supplier to determine possible gauge thicknesses
Monolayer polypropylene film	H	S	Needs testing to determine film's tolerance to manufacturing, distribution, and consumer use	Supplier to determine if a single polypropylene film exists that has properties equal to both of the current polypropylene films
Explore alternate materials to polypropylene	M	M	Current film supplier makes polyester films in-house	Supplier to cost out polyester option
Explore different suppliers of polypropylene	M	M	Using different polypropylene requires testing at film supplier	Supplier to contact polypropylene suppliers
Explore savings opportunities at polypropylene suppliers	L	L	May require testing at polypropylene suppliers	Supplier and purchasing to contact polypropylene suppliers to determine if opportunity exists for a future project
Surface print vs. reverse print	L	M	No cost savings, may experience ink pick with surface print	None

Table 3.1. *(continued).*

Idea (top ideas in italics)	POS	Timing	Comments	Next Steps
Narrow web width	L	S	Cost savings minimal, could cause production issues	None
Larger rolls of film	L	L	Weight of larger rolls exceeds current equipment's capacity and OSHA concerns	None
Cold seal vs. heat seal	L	L	Cold seal does not provide hermetic seal	None
Purchase quantity and inventory opportunities	L	L	Opportunity explored last year	None
Decrease graphics ink coverage	L	S	Minimal savings	None
Savings in the film making process	L	L	Supplier installed new equipment two years ago for savings	None
Produce film in-line	L	L	Lead time too long, space and printing issues	Consider for long-term research project
Print graphics in-line	L	L	Lead time too long, space and odor issues	Consider for long-term research project

Will the graphics be of the same quality if printed on a thinner
gauge film?
Will there be a visible difference between a thinner film and the
current film?
• not affect the product's quality perception with consumers
Will a thinner film be noticeably different to a consumer in feel,
look, or performance?
• be completed prior to a manufacturing cost increase (18 months)
What is the timing for qualifying a new film?

Bill cannot answer all the questions he has generated. So he asks himself, *"What information do I need to know to answer these questions?"*

Bill realizes he needs to know how film cost is related to the film's thickness. He also needs to understand if one or both layers can be made thinner, and what concerns may be associated with doing so. Does the film supplier have any limitations on producing thinner film? Are there other grades of polypropylene whose properties are maintained at thinner gauges? Bill wonders if there are any similar products to Credo's that are in thinner gauge film and whether or not they are packaged in polypropylene.

Bill next asks, *"Who has information that can help me?"* Bill decides to call the film supplier to discuss the options for a thinner gauge film, potential concerns, timing requirements and to request price quotes. He also decides to meet with Katelynn, a senior packaging engineer in his department, to get her input because she has a lot of film experience. He considers calling producers of polypropylene directly but decides to wait until he has more information.

Finally, Bill asks, *"Where are other resources located that may have more information?"* Bill feels he needs to know more about polypropylene film and the various grades. He does not understand how the grades of polypropylene vary on a molecular level and how that affects end properties of the film. He decides to check the textbooks in the packaging resource area in his department and the packaging and polymer periodicals for any applicable articles. To gain a better understanding of the manufacturers of polypropylene, he makes a note to look in his supplier resource book. He also wants to look through a file of literature from polymer suppliers that he has been collecting. Bill plans to go to the store and purchase competitive product and identify those packaging films and measure their gauge. He also notes that he needs to make a list of the Credo's package performance criteria so he will be able to evaluate test films versus the current film.

Bill will complete this process for each of the top brainstorm ideas.

Step 2: Determine the Probability for Success

Bill gathers all the information he can and then determines how well each idea will satisfy the criteria for success.

Several weeks have passed during which Bill and his film supplier have determined options for the idea of decreasing the gauge of one or both layers of polypropylene. Based upon the supplier's production processes and their internal lab tests, they suggest decreasing the gauge of the film from 148 mil to 115 mil by making each of the two polypropylene layers thinner. Based on the sales volume of Credo's, the estimated annual film savings is $1.4 MM. An alternate material that maintains its properties at thinner gauges is also suggested. No additional savings are found but concerns over packaging performance are less. The supplier has also produced samples of graphics printed on thinner polypropylene film that look identical to the current Credo's package.

Bill will not be able to fully answer questions regarding package performance without actually testing packages produced with a thinner gauge film versus the current film. Lab test data regarding barrier properties, optical properties, and mechanical properties indicate that a thinner gauge film should be able to meet the Credo's package performance criteria. Bill has also learned that one of Credo's competitors is currently using a package of 110 mil polypropylene film. Initial thoughts are that consumers will not notice a change to a thinner gauge film, but a consumer test will be part of the qualification process.

Based on the information Bill has gathered, he feels confident that a thinner gauge package can be qualified within the specified time frame. The cost savings, although not $2.8 MM, is still significant. He rates the probability of success for the idea of decreasing the two layers of polypropylene as high (H).

Bill again completes this process for each of the top ideas that have been selected for further review.

Step 3: Reevaluate Ideas

Bill has now gathered information on the top brainstorm ideas. He identified five options that have a high probability of success with cost savings ranging from $1.4 MM to $2.1 MM. The team reviews the original brainstorm list and feels they are pursuing the right ideas even though the estimated film cost savings will not meet the targeted $2.8 MM.

Step 4: Determine if More Idea Generation Is Needed

The team decides not to conduct more idea generation sessions. They feel there may be savings in other aspects of the production process and will suggest to management that additional savings opportunities be identified and pursued beyond the packaging film.

Step 5: Determine What Next Steps Are Needed to Fully Qualify (i.e., Proving Functionality) the Selected Idea or Ideas

Now that the team identified film options, they can complete a more detailed project timeline.

Scenario B: Oula Package Redesign

Susan compiles all of the package redesign ideas into a memo and distributes it to her team members. Some of the ideas will require product development. A portion of the document is shown in Table 3.2.

Susan now begins the process of evaluating the top package redesign ideas using the five steps for determining feasibility.

Idea: Stacking Package: Two Circular, Glass, or Plastic Containers

One container has a screw-on closure, and its base has lugs so that it screws onto the other container to act as its closure. When fitted together, the containers appear to be in a stack of two.

Step 1: Determine if the Idea Meets the Criteria for Success

Critical Success Factors (CSF) (from the team charter):
- The redesigned package must be rated by consumers as a

Table 3.2. *Oula Package Redesign Project: Brainstorm Ideas.*

Idea (top ideas in italic)	POS	Timing	Comments	Next Steps
Stacking Package: two containers that screw together, each with its own closure	H	S (stock) L (custom)	Consumer can replace one container when empty, creams won't mix	Investigate stock container and custom design possibilities
Side-by-Side Package: two containers attached at sides, each with its own closure	H	S (stock) L (custom)	Consumer can replace one container when empty, creams won't mix	Investigate stock container and custom design possibilities
Two Tubes Package: two tubes that fasten together	H	S	Consumer can replace one container when empty, creams won't mix	Contact tube supplier to discuss idea
Dial out moisturizer	M–H	M	Use a similar package from the men's line that dispenses through slits in package with turn of a dial	Research men's line package for feasibility
Pump dispensers	M	M	Consumer data indicates poor liking ratings for pump packages	Determine if an improved pump system is available
Creams in a stick	M	L	"Paint" cream on using the tube, will require some product reformulating	Discuss idea with product development
Spray package for one of the creams	M	L	Spray on base cream	Discuss idea with product development

39

significant improvement versus the current package for closure integrity and product containment.

- The redesigned package must be rated by consumers as an improvement versus the current package for overall liking of the packaging.
- The redesigned package must protect the product at parity to current.
- Management agrees to approve up to a 10% packaging material cost increase given the redesign meets all CSF.
- Additional costs for tooling or equipment may be acceptable given the redesigned package meets all CSF.
- Management approval for a redesign needed by 6/1 in order to obtain any funding from this year's capital budget.

Susan realizes she needs additional information to determine if the Stacking Package meets the criteria for success. Thus, she asks herself the following questions as she prepares to gather more information.

She first asks, *"What information do I need to know to answer this question?"* Information regarding consumers' liking for the Stacking Package cannot be obtained until consumer tests are conducted. The feasibility assessment must be done prior to consumer testing for Susan's project. Susan can assess consumer response to the package, but she would like some information to substantiate a recommendation to proceed with the idea. Consumers' thoughts and attitudes from prior consumer tests regarding glass versus plastic for cosmetic and toiletry packaging would be helpful.

Susan also needs package cost information. She must first determine if there is a stock container that will work for the Stacking Package idea or if a custom container is needed. Susan needs to know what is available in glass or plastic stock containers and what companies to talk to regarding custom containers. If a custom container is needed, how much will tooling cost? What will be the packaging production requirements? What new packaging equipment will be needed, and how much will it cost? Susan is also concerned that a glass package will result in product damage during distribution and cause production concerns.

Susan next asks, *"Who has information that can help me?"* Chris in marketing has copies of past consumer tests. Several of the other packaging engineers utilize glass for their products and might have some useful information regarding consumers' liking for glass as well as product protection concerns, especially production and distribution issues. Susan has a list of glass and plastic container suppliers but also decides to phone Mark in purchasing for his recommendation on which suppliers can help her in her container search. She also decides to ask her packaging peers for supplier contacts that specialize in custom design packaging. Susan needs to talk to Paul in operations for his input on packaging equipment requirements for this style package.

Finally, she asks, *"Where are other resources located that may have more information?"* Susan makes a note to ask one of the glass suppliers for a tour of their facility so she can learn more about glass containers. She plans to travel to the manufacturing facility to view glass and plastic packages on the manufacturing line and to talk with the line supervisor for his input. Susan also wants to learn more about the competition's packaging outside of Oula's target market segment. To do this, she plans to take some field trips to see how the competition's products are displayed and to purchase some samples.

Susan completes this process for each of the top packaging ideas. She compartmentalizes all of the questions that need answering next to each person she needs to contact and begins making phone calls.

Step 2: Determine Probability for Success

After six weeks, Susan has determined that a stock container will not work for the Stacking Package idea. Her research indicates that a plastic container is preferred by the manufacturing facility and will be less expensive to produce. She has narrowed the focus to two suppliers after meeting with several to discuss a custom package. Each provided Susan with several detailed drawings of proposed package designs and costs for tool production and packaging materials. Package cost will increase 5.5%. A prototype tool will cost $29,000, and the production tool will total another $280,000. There is the possibility of amortizing the production tool cost over the first three years of package purchases.

Some adjustments would be required for the package to be produced on current production equipment. The engineer at the manufacturing facility estimates $250,000 to $300,000 per packaging line will be needed to make the necessary additions and changes.

At this point in the project, Susan makes an assessment regarding consumer reaction to the Stacking Package. The screw-on closures will not pop open like the current package's flip top. The two creams will be accessed separately, eliminating the issue of one cream spilling over into the other. An empty container can be replaced with a full one, eliminating the issue of not finishing the two creams at the same time. Consumers like having the two creams together and this would not change. Aesthetically, Susan feels the new package would be a great improvement over the current package. Features such as transparency, color, shape, and weight can be altered for marketing appeal. A more upscale appearance coupled with the added features should make the package an improvement in the eyes of the consumer. All of this is Susan's opinion. She has reviewed all of the applicable consumer data that is available and finds that consumers like glass and plastic packages as long as they are not too large or heavy and that both packages can convey a sense of quality.

Susan feels the team can create concept boards or mock-up package samples for consumer concept tests by 4/10. Thus, the consumer concept test can be completed and a management approval meeting held prior to the due date of 6/1. The data from the concept test will be critical in determining if the Stacking Package idea should be pursued given the projected tooling and equipment expenses. Susan feels strongly that the Stacking Package idea should be concept tested. If concept test results are extremely favorable, the project's financial requirement may be agreeable to management.

Susan has gathered similar information for the two other top brainstorm ideas. She presents the information to her fellow team members.

Step 3: Reevaluate Ideas

The team reviews all of the information that Susan gathered and makes some minor changes to the design of some of the packages. They decide to submit four concept boards for consumer concept testing. Two Stacking Package concepts, a Side-by-Side Package concept, and a Two Tubes Package concept are submitted.

Step 4: Determine if More Idea Generation Is Needed

At this time, the team will not brainstorm for more ideas. They may decide additional ideas are needed pending consumer concept test results.

Step 5: Determine What Next Steps Are Needed to Fully Qualify (i.e., Proving Functionality) the Selected Idea or Ideas

The team creates a rough timeline for each of the four package concepts that are being consumer tested. This will give them an understanding of the major steps and time commitment needed for each package idea. Once the consumer concept test data is available, they will have a better idea of what package concept to pursue and will then create a detailed project timeline for that concept.

Scenario C: Carton Crisis

Kathy receives the cartons the next day and takes them directly to the lab. The coefficient of friction test results from lot KL510 cartons is 0.35 kinetic, while lot KL812 is 0.62. Kathy notes that the carton surfaces also feel different. As she is finishing up the tests, Bob phones with the carton manufacturing information.

The coating material used on the cartons produced in lot KL812 was purchased from a different supplier than is normally used. The current coating supplier had not been able to deliver their order and was expecting to be six weeks late. Bob explained that they had used an alternate coating supplier who produced a replica of the specified coating. They had processed the coating the same way and had not had any issues on their end in producing the cartons. Bob explains that this was the first he was finding out about the coating supplier change, and that it is not normally their policy to make such changes without Kathy's permission. Unfortunately, their current coating supplier is anticipating more delays in future coating deliveries. They were able to produce a limited quantity of cartons using the specified coating supplier and are shipping them later in the day to Kathy's manufacturing facility.

Kathy shares the coefficient of friction data with Bob and expresses her concerns with the alternate coating. The situation is unacceptable, and a resolution must be found quickly. Bob agrees but needs some time to work out a solution with his technical staff. He promises to contact Kathy the next day with their action plan to find an alternate coating supplier.

CHAPTER 4

Consumer Testing for Packaging

CONSUMER TESTING IS a useful, though expensive, tool; because of test sample size and consumers' ever changing opinions and thoughts, it is also a nonexact science. Very small segments of the total marketplace are tested. Thus, the margin for error in the results can be large. The expense to test a larger segment for more statistically valid data often does not warrant the test. Even with its limitations, consumer test data is a very necessary guidance tool for making decisions regarding the direction of many packaging projects.

Consumer testing should be conducted by people who are trained in the field. Expertise is needed to choose the right test method, to select the correct consumer audience, to properly develop and disperse the test, and to correctly interpret the data. Some large companies have their own in-house consumer test experts who design consumer tests to fit a packaging engineer's project needs. There are also many independent companies specializing in consumer testing that can be hired to perform the same duties.

There are many methods for testing package concepts with consumers. Depending on the project, it may be necessary to conduct several consumer tests prior to making a final decision on package design. A first step may be to consumer test many package concepts and then proceed with the more in-depth consumer usage test using fewer packages.

CONSUMER CONCEPT TESTING

Consumer concept testing is conducted using photos, drawings, or mock-up samples of the package concepts. The drawings or photos are often called boards because they are usually mounted on foamcore or paperboard for presentation. Mock-up packages may be made out of wood or plaster to simulate the final shape and design of the package concept. The mock-up packages are nonfunctioning representations of the actual package concepts. Whether using boards or mock-ups, the package features (such as a reclosable or a dispensing feature) usually require some explanation for the test participants.

Consumer concept testing is used when there are many new and innovative packaging ideas being considered, particularly for growth and development projects. It is quite useful in narrowing the list of packaging concepts, and new information is often obtained regarding modifications to the concepts or even additional package ideas not previously thought of. All of this is done without incurring a lot of package development expense. Consumer concept testing is usually much faster and less expensive than consumer usage testing.

It is not always necessary to conduct consumer concept testing. If a project warrants consumer testing, but the scope of the project entails a minor package modification or few possible packaging options exist, then it is best to skip consumer concept testing and go directly to consumer usage testing.

DEVELOP PACKAGE PROTOTYPES

Prototype packages are packages that look, feel, and function just as a final package is intended. The only exception may be the package graphics. Adhering labels to the packages rather than having the packaging materials printed may be done to save money and time. It may seem obvious to state that the package features must function and package variability must be kept at a minimum, but when using prototype packages that have been handmade or produced on prototype equipment, it can be difficult to control all of the variables.

The development of package prototypes is rather simple if the packages can be produced on current packaging equipment in the manufacturing facility. New packaging materials or graphics may be needed, but the process to make them is in place. This is often not the case. Usually, additional packaging equipment is required, often equipment that is unfamiliar to the project team or perhaps equipment that is experimental. The product may have special manufacturing needs that can further complicate the process of making prototypes. A small number of package prototypes are usually needed for consumer testing, 50 to several hundred for each package concept being tested. The number may be too few to produce on large-scale production equipment, but too many to make by hand.

When determining the best method for prototype production, the following should be considered:

- specific features of each prototype package design
- number of each prototype needed
- packaging and product manufacturing environment requirements
- cost of the various methods of prototype production
- timing

Sometimes packages can be produced by hand or by using simple lab equipment and a lot of manual labor. This method is usually the least expensive but the most time consuming. The potential for product contamination is high when making package prototypes by hand. Using manufacturing equipment will decrease the production time for the prototypes, but the time needed to obtain new equipment can be very lengthy, especially if any custom work is required. Some packaging equipment suppliers have smaller versions of their packaging machinery that they will loan or sell for prototype production. Some of these units can even be modified to meet special packaging needs.

CONSUMER USAGE TESTING

In consumer usage testing, consumers get the opportunity to actually use a package and make comments. It is customary to use prototype packages with full graphics for consumer usage testing. Each consumer participating in the study will receive his/her own package to evaluate.

Information is gathered on how well the package features actually function and what consumers' opinions and feelings are about each package tested. There are a number of methods for conducting consumer usage tests. Sometimes it will be important for consumers to use the package for an extended period of time, and in other instances, five or ten minutes of consumer interaction can provide the needed information.

CONSUMER PACKAGING TEST METHODS

There are many test methods used for consumer concept and consumer usage packaging tests. Before selecting a method, the following must be determined:

- *a clear test objective:* What one question must the data answer?
- *other information that must be obtained for the results to be meaningful*—such as the following: Will consumers pay more for the preferred product/package system? Do consumers feel their usage will increase with the package change? Do consumers feel their needs are no longer met if the package is modified?
- *the consumer profile that is to be tested:* Consumers who are loyal, heavy product users; consumers of competitive products; consumers who switch from one brand to another; or a mix? Sometimes a variety of consumer groups are tested to learn how different segments will be affected by the packaging change. The project goal will guide the consumer profile selection.

Prior to the consumer test a prescreen is done to determine if a person qualifies to participate. The prescreen is a series of demographic and product specific questions asked to potential participants to verify that they fit the desired consumer profile.

Types of Consumer Tests

Mall Intercept

The mall intercept consumer test is performed in a shopping mall or shopping center. People are approached while they are shopping and asked if they will partake in a consumer test. If they agree, they are asked a few prescreen questions to ensure they qualify. Tests are given in the mall arena or consumers are led to a nearby room and presented with the packaging test.

Phone Interview

The phone interview is used for prescreening and for asking follow-up questions regarding consumer tests that have been completed. Phone interviews can also be used to conduct the consumer test but only if other materials such as photos or packaging prototypes have been sent to the consumer being interviewed.

Individual Interviews

Consumers can be interviewed individually regarding a package, but this can be an expensive and time-consuming process. Sometimes, consumers will be seated together in a room, but rather than speak to each other, they will view packages on their own and give feedback via a written survey.

Focus Groups

Participants are brought together at a test facility and asked to discuss a package or packages as a group. A two-way mirror may be used to observe or videotape consumers. Prescreening for focus groups is done via telephone or mail.

In-Home

Packages are used by consumers in their homes for this consumer test. Sometimes, consumers are asked to use the product in a specific way or by a specific date. The package is either mailed to the consumer, or the consumer picks it up at a predetermined location. Participants may be asked to keep a diary regarding their use of and thoughts about the package. Once the allotted test time is over, consumers are asked to provide feedback by either filling out a survey that is to be mailed back to the test givers, participating in a phone interview, or by participating in a focus group discussion.

Test Market

A test market is a very large-scale usage test in which only one package is tested in one market region for an extended period of time. The test is accompanied by advertising and other promotional support. These tests normally last several months to a year and are quite expensive. Test markets are done when more substantial data is needed to support the desired packaging change and/or to justify a project's high financial commitment.

Testing Methodology

Consumer tests need to be presented in a manner that will not influence or bias the test results. The sequence in which the packages are presented is rotated among participants if each participant is to evaluate all package variables. For instance, if there are four packages labeled "A," "B," "C," and "D," one participant would view them in that order, another would see package "B" first, then "C," "D," and "A," and another would see "C," "D," "A," and "B," and so on. The same rotation method is sometimes used for survey questions to ensure that the results are not affected by the order in which the questions are asked or by test fatigue.

For some tests, participants may only evaluate one or two package variables even though there are more in the study. This is done when the package evaluation process requires high involvement on the part of the test participant, such that they would become disinterested or overwhelmed if they had to evaluate all of the test variables.

For some studies, consumers are asked to compare and contrast one package versus another. This is called hedonic test methodology. For other consumer tests, the consumer answers the same set of questions for each package variable being evaluated and rates each package individually. That process is called sequential monadic.

CONSUMER TESTING FOR PACKAGING:
CHAPTER SUMMARY

- Consumer concept testing is conducted using photos, drawings, or mock-up samples of the packaging concepts. It is used when many new and innovative packaging ideas are being considered.
- Prototype packages are packages that look, feel, and function just as a final package is intended.
- Consumer usage testing is conducted using prototype packages to gather information on how well the package features actually function and to obtain consumers' opinions and feelings about each package variable.
- A prescreen is a series of demographic and product-specific questions potential test participants are asked in order to determine if they fit the desired consumer profile.
- Types of consumer tests are mall intercept, focus groups, in-home, individual interviews, phone interviews, and test market.

PACKAGING PROJECT EXAMPLES

Scenario A: Credo's Productivity Project

Due to the nature of Bill's productivity project, the consumer testing will be conducted as part of the Proving Functionality Phase. The team does not feel that it would be a wise decision to wait for consumer test data results before proceeding with the rest of the proving functionality tests. Various packaging films are to be used, but no manufacturing equipment changes are required. Thus, it makes sense to conduct one plant trial to produce all of the packages needed for the consumer test and the proving functionality tests.

The objective for the consumer test is to determine if consumers notice a difference between the current Credo's package and the thinner gauge productivity package options; and if so, do they view the difference as a negative? The team needs to feel confident that a thinner gauge film will not affect the consumer's perception of Credo's product quality.

Marge in consumer behavior specifically defines a thinner gauge package as acceptable if consumers rate quality at parity or better to current and if they rate the overall package at parity or better. Prototype packages can be easily attained, thus, there is no need to conduct a concept test. The team views the productivity packaging change as very low risk. Marge indicates that a mall intercept test given in one city would be best for the study. Consumers who meet the prescreen requirements will be taken to a separate room in the mall and shown one package at a time in a sequential monadic manner. Consumers will be asked to rate a series of attributes on a scale of one to five for each package.

Scenario B: Oula Package Redesign

Concept Test

Current Oula consumers are dissatisfied with the package closure and product containment features. Marketing feels that Oula's volume growth has not reached its full potential due to some consumers rejecting the product because of the inconvenient package. Thus, Barb feels both Oula product users and nonusers must participate in the consumer test. She will conduct individual interviews with consumers using boards to gain consumer information on the new package ideas. Each consumer is to be shown the four concepts, one at a time in varying order. There are two variations on the Stacking Package idea, one Side-by-Side Package and one Two Tubes Package. Additional questions will be asked to learn consumers' thoughts on glass versus plastic and tubes versus containers.

After completing the concept test, the data shows that the two Stacking Package ideas rate highest for convenience of use, quality perception, closure integrity, and product containment. Both were viewed as significant improvements versus the current package. The other two concepts were seen as improvements versus the current package but not as significant improvements. The consumer data also indicate that a plastic package will meet the needs of the consumer. Plastic will also be less expensive. The team finished all of their tests and information searches two weeks prior to the 6/1 deadline.

The next step is to conduct usage testing. Prototype packages will be needed and in order to secure the necessary funds for the prototypes, management's support will be necessary. The team presents the results of the consumer concept test (see below) to management with a recommendation to usage test both Stacking Package ideas. Due to the impressive concept test results, management agrees with the recommendation.

Prototype Production and Testing

Both of the Stacking Package ideas involve two separate containers, each with its own closure. Susan required that the containers be designed for a standard closure size. Thus, she can obtain stock closures for the packages, saving on the time and expense required to produce a

Concept Test: Overall Package Liking (9-point scale)		
Package	**Oula Users**	**Oula Nonusers**
Current	4.9	4.8
Stacking Package A	6.9	7.0
Stacking Package B	6.8	6.7
Side-by-Side Package	5.4	5.2
Two Tubes Package	5.3	5.2

closure tool. One of the other things Susan has done in the design phase is determine the secondary and tertiary packaging. The finished Stacking Package will be placed in a carton and then into a shipping case. Susan calculated the carton size and the shipping case configuration to ensure that she efficiently maximized pallet space.

Susan double-checks all of the measurements on the blueprints for each package design before she approves the finished package drawings. She submits the signed drawings to the supplier that will produce the unit tools. The packages will be injection molded. The supplier will need three months to complete the unit tools for both packages. Once completed, they will produce several packages and send them to Susan. She will verify the dimensions of each package using the optical comparator, paying particularly close attention to the closure thread dimensions. With a micrometer, she will measure the wall thickness of each container in quarter-inch increments around the packages and up and down the package. She will also verify the overflow capacity of each container and measure the torque required to remove the closures. Susan's supplier does a good job of providing her with packages that meet the specified dimensional requirements. She is able to approve both unit tools.

While the unit tools were being produced, Susan worked with the manufacturing facility to ensure that the Stacking Packages could be filled at the plant for the consumer usage test. Change parts for the Oula product filling and capping systems were needed. No other packaging equipment changes will be made at this point in the project. For now, all packages will be manually loaded and unloaded from the line. Once the filling system is ready, Susan obtains enough of each package for all her test needs and schedules a plant trip.

Usage Test

Barb once again screens for Oula users and nonusers for the usage test. An in-home consumer test is planned. Some of the test participants may be asked to partake in focus group discussions if additional information is needed. Each participant is asked to use all three packages: the current package and the two Stacking Package concepts. Each is sent one package and asked to use it for one week, keeping a diary of their likes and dislikes. At the end of the week, they are to complete a survey and mail it back to the consumer testing facility. Another package will then arrive to be used for the next week, and finally, the third and final package will be sent. The order that the participants analyze the packages is varied.

Survey results are tabulated. The diaries provide some additional insights. Both of the Stacking Package ideas are again seen as significant improvements versus the current package, but one of the ideas has a stronger liking rating than the other. Oula users and nonusers liked the same Stacking Package. The team is concerned that consumers may not understand the container replacement process because they only viewed the packages for one week each. It is decided to conduct a focus group to further explore the situation.

The focus group indicates there may be some confusion regarding replacement of one of the two stacking containers when empty. The focus group participants are asked for their input to simplify the replacement process. Drawings and specific wording to explain the process on the carton are suggested by the focus group participants. They also feel it would be helpful if each one of the two containers in the Stacking Package is a different color.

The team agrees to incorporate the ideas from the focus group and feels confident that the Stacking Package idea should be pursued. They now need to present their recommendation to management.

Scenario C: Carton Crisis

Consumer testing is not required for the carton situation.

CHAPTER 5

Final Concept Evaluation

THE LAST STEP in the planning phase is to use all available data to evaluate the chosen package concept and, as a team, to recommend to proceed or not. Many of the original packaging ideas will have been eliminated through feasibility and consumer testing, so that now there are only a few that could satisfy the project goal. The decision drivers for selecting a package are technical feasibility, consumer preference, time requirement, and financial commitment. If the risk associated with the packaging change is low, and finances and time are not a concern, two or three packages may be selected for further evaluation.

The packaging engineer is responsible for providing most of the information needed to make a decision on which concept to pursue. Team members and suppliers are relied upon for input, but it is the packaging engineer's responsibility to secure all of the necessary information. Confidence in the information comes from conducting thorough research and utilizing field experts. As a safeguard, a project review can be held at this point in the package development process.

Through feasibility testing, data are obtained to determine technical merit. By working together with consumer test experts, consumer preference is evaluated. Estimating the time needed to prove functionality and launch a package concept requires thinking through all the processes involved in qualifying the package concept and allotting the correct amount of time for each. If new packaging equipment or a production tool is required, the project lead times can be quite long.

Much time and effort are needed to predict the true financial requirement for a packaging project, especially growth initiatives which usually require packaging material and equipment changes. Both capital and operating expenses must be estimated. A capital expense is a tangible long-term asset such as packaging equipment. Operating expenses are those items required for the manufacturing facility to operate, such as labor and benefits, material and supplies, utilities and service, and other operating costs like depreciation and insurance. If capital estimates are low, the team may have to scramble to extradite more funds later in the project. Portraying a true financial picture is extremely important to understanding how packaging changes can affect a product line's monetary contribution to an organization.

The packaging engineer participates by providing some of the financial information, the packaging material cost, and capital expenses, but it is usually marketing that calculates whether the project is financially feasible. A per unit packaging material cost is given to marketing so the standard product unit cost (SPUC) can be calculated. Product cost and other variable costs such as labor and manufacturing are also considered when determining SPUC. There may be one or several suppliers that can provide packaging materials needed for a project. Quotes are obtained based upon projected package sales volumes, package dimensions, and printing requirements. In many organizations, purchasing will negotiate final material costs and should be involved in obtaining quotes.

If capital funds are required for equipment, the packaging engineer or appropriate team member works with suppliers to specify needs and design accordingly. The suppliers will then provide cost quotes. At this point in the project, no official commitments are made. If the project proceeds, purchasing will negotiate a final cost and make contractual agreements based upon an approved design. Thus, purchasing should view all quotes to ensure that additional costs and concerns have not been omitted. Capital expenses are not part of SPUC, but marketing is often in charge of allocating the available capital dollars.

By means of a team, a document including all of the technical, consumer, timing, and financial data to support the team's recommendation is compiled and then presented to a decision-making group, usually consisting of management and senior personnel. Included in the document and presentation should be the other packaging ideas that were considered and the reasons for not pursuing them. Samples of all the package concepts should be available at the meeting. Some smaller projects will not warrant a formal presentation. A discussion with a manager of the findings to date and the next steps may be all that is needed. Data documentation is always recommended, regardless of a project's scope.

GO OR STOP DECISION

If a project is well thought out from the beginning with a clear goal and specific criteria for success, the decision to proceed will be much cleaner. A "go" or "stop" decision may be given immediately, or additional information and more time to consider all of the data may be requested. Financial resources are usually required for a project to proceed beyond the planning stage. Some projects may be put on hold until the organization's financial situation changes, or limited funds may be allocated so that a portion of the proving functionality tests can be started.

FINAL CONCEPT EVALUATION: CHAPTER SUMMARY

- The decision drivers used to evaluate a package concept are technical feasibility, consumer preference, time requirement, and financial commitment.
- A capital expense is a tangible long-term asset such as packaging equipment.
- Operating expenses are those items required for the manufacturing facility to operate, such as labor and benefits, material and supplies, utilities and service, and other operating costs like depreciation and insurance.
- Standard product unit cost (SPUC) includes packaging material, product, and other variable costs such as labor and manufacturing.

PACKAGING PROJECT EXAMPLES

Scenario A: Credo's Productivity Project

Bill has completed the feasibility research on the five productivity film options that were identified. He has compiled the information in a document and is presenting it to his team members and managers. He begins his presentation with a summary page of the film options.

Credo's Productivity Project: Summary of Film Options

Film Options	Film Gauge	Est. Annual Savings	POS	Comments
A. monolayer pp	1.15 mil	$2.0 MM	H	competition using monolayer pp film
B. alternate material, monolayer	1.10 mil	$1.9 MM	M–H	shelf life required
C. decreased pp in multilayer film, alternate pp supplier	1.15 mil	$1.6 MM	H	alternate pp supplier used by competition
D. decreased pp in multilayer film	1.15 mil	$1.4 MM	H	no issues expected
E. alternate material, multilayer film	1.20 mil	$1.4 MM	M–H	shelf life required

Bill explains the structures of each of the films and any potential issues that might be expected. He has rated two of the films as being medium-high for probability for success due to the film being a material that has never been tested. Credo's has always been packaged with polypropylene. Bill then presents the test plan and the project timeline. Several of the film options will likely be eliminated once the first portion of the tests are completed. He will then conduct an extended trial with one or two chosen films.

Some discussion takes place, but the group feels confident that the five options are worthy of pursuit. Everyone agrees with the plan, and approval is given to move forward with the Proving Functionality Phase. A request is made to meet again prior to the extended trials.

Scenario B: Oula Package Redesign

The team gathers again to present the usage test data with hopes of gaining approval to continue the project. The team starts by presenting the usage test results that indicate an overwhelmingly positive consumer response to the Stacking Package A concept. They also provide additional support information to back their recommendation of pursuing Stacking Package A.

Usage Test: Overall Package Liking (9-point scale)		
Package	**Oula Users**	**Oula Nonusers**
current	4.7	4.8
Stacking Package A	7.4	7.2
Stacking Package B	6.3	6.0

Recommendation: Stacking Package A

Meets all criteria for success:

- viewed as a significant improvement versus current for closure integrity, product containment, and overall package liking
- package material cost to increase 5.5% (CFS is 10% max.)
- tooling expense: $280,000
- equipment expense: $650,000
- product protection at parity to current: product/package interaction stable (no resin change), improved distribution protection (lab distribution test data)
- all tests have been completed prior to the 6/1 date to obtain funding from capital budget

The current Oula SPUC is $5.95. Marketing has calculated the SPUC for "Stacking Package A" to be $6.95. The Oula dual moisturizing product is currently priced lower than its two direct competitors. Consumers indicated in the usage test that a price of $6.95 was reasonable.

The management team is very impressed with the Oula project results. Capital budget funds have been tentatively set aside for the project. A final review of the budget will be required prior to project approval, but the consensus is that the project has a high likelihood of obtaining approval.

Scenario C: Carton Crisis

Kathy meets with her manager to discuss what she has discovered to date on the defective cartons. She has organized the data on one page to help with the discussion.

Kathy Malani, Sept. 4

Project: Defective Carton Issue

Situation: An alternate coating supplier was used due to supply shortage of current coating. Replacing the current coating at alternate supplier resulted in COF changes causing cartons to jam in the packaging machine. Coating supply concerns are expected to continue, thus an alternate coating supplier must be qualified.

Data:

Current Carton COF (KL510):	0.35 kinetic
Defective Carton COF (KL812):	0.62 kinetic

Next Steps:

- Carton supplier to analyze coating used on cartons KL510 and KL812 to determine chemical differences (Sept. 6).
- Carton supplier to find a coating supplier that can replicate current coating (Sept. 30).
- Samples of cartons with new coating provided to Kathy for lab tests (Oct. 10).
- Carton trial at manufacturing facility (week of Oct. 14).
- Lab distribution testing (week of Oct. 21).
- Carton specification modified to include two approved coating suppliers and COF requirement (Oct. 30).
- Kathy to work with carton supplier to determine acceptable COF target and range (Oct. 30).

Kathy uses the page she has created to update her manager. They discuss in more detail the lab tests Kathy plans to conduct and the process she plans to use to specify the coefficient of friction (COF). Kathy's manager suggests that glue adhesion also be evaluated. They next discuss the current carton supply situation. Kathy explains that the carton supplier can meet their present carton needs as long as a new coating is sourced and approved by October 30. Purchasing has made it clear that the carton supplier will be financially accountable for any production inefficiencies at the manufacturing facility if they do not meet the October 30 date. Kathy hopes that will be motivation for the carton supplier to hit the target date, otherwise they will have a situation on their hands. Kathy's manager agrees with Kathy's approach to the project so far and asks that she continue to be kept posted. She then congratulates Kathy on her good work.

CHAPTER 6

Proving Functionality

PACKAGE TESTING

PROVING FUNCTIONALITY OR qualifying a package means to ensure that every detail has been tested and thoroughly considered so that a package complies to all standards and functions in all environments to which it may be exposed. The objective is to strive for a flawless execution of the new package. Lab tests, plant trials, and thorough exploration of any potential issues, such as quality or legal, are all ways of proving the functionality of a package.

Developing a Test Plan

To establish a test plan for package approval, follow these four steps:

(1) Determine the package performance criteria.
(2) Determine the tests that must be completed for package approval.
(3) Determine what action(s) must be taken for each test.
(4) Determine timing for each test.

Tests are performed on the packaging material and on finished packages. Sometimes a feature cannot be thoroughly tested due to timing, cost, or technical constraints. If this is the case, assigning a confidence level to the test data that has been collected will indicate the level of assurance that the performance criteria have been met. All data and information, including any weakness in the test plan, must be exposed so that an informed team decision can be made when determining whether or not to pursue a packaging idea.

Packaging Material Testing

Package material test data are useful for comparing several materials and are good indicators of the end performance of the final package. Some package material tests may be performed as part of the feasibility assessment process. Material tests are also used as quality control by material suppliers.

There are three properties of packaging materials:

(1) *barrier properties:*	gases (oxygen, carbon dioxide, nitrogen)
	water vapor (moisture)
	flavor
	light
(2) *optical properties:*	transparency
	opacity
	gloss
	haze
(3) *mechanical properties:*	dimensional stability
	high temperature stability
	low temperature stability
	flex crack resistance
	heat sealability
	stiffness
	impact strength
	puncture resistance
	tensile strength
	elongation
	tear strength
	coefficient of friction
	antistatic
	dead fold
	hot tack

Standard test procedures for testing material properties can be found in ASTM and TAPPI journals. There may also be circumstances in which a packaging engineer will develop his/her own material test method. For instance, if a finished package is experiencing film fracture, but falling dart test data indicate good puncture resistance, then a material test that is more sensitive and that better replicates the end use of the film may be required. The film supplier can also use the test as a standard when manufacturing the film.

Finished Package Testing

The finished package tests are determined by the requirements of the product and the environment to which the package is subjected. There are some standardized tests for finished packages, simulated distribution testing and leak testing for example, that can be found in ASTM and TAPPI journals. Many finished package tests though, will be developed specifically for a particular package and its unique parameters.

Products often require a package to protect against spoilage due to oxygen, water, flavor, or light. Material barrier testing will be a good indicator of product protection, but to fully qualify a packaging material, it must be tested with the product in the finished package form. To verify that a package can provide the required protection, shelf life testing is completed. Shelf life testing is done by using the finished package and product and testing them in the intended storage environment over the life of the product. The product and package are analyzed at specific time intervals depending on the particular concerns. For example, packages might be evaluated for oxygen ingress or for seal strength over time. The product may be tasted by a panel of product experts for off-flavor, freshness, etc., or be measured for mold growth or discoloration. Testing frequency will be somewhat product specific, daily, weekly, monthly, etc. Testing outside normal storage conditions may be done to understand what effect it has on product spoilage. To obtain finished packages for shelf life, the best situation is to package products using the packaging manufacturing equipment at the manufacturing facility. To minimize any test variance, all test and control packages should be filled within the same time frame using product from the same lot or batch.

Accelerated shelf life testing can be conducted for some products. Through testing and experimentation, it may be determined that exposure to extreme conditions for a shortened time period models the effects of exposure to normal storage conditions over normal product shelf life. For many food products actual shelf life testing must be performed to ensure product safety.

Most packages are subjected to a variety of environments (Figure 6.1), and all must be considered when qualifying a package. The envi-

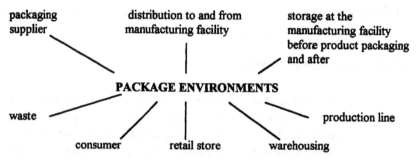

Figure 6.1. Package environments.

ronment may also change throughout the year due to seasonal changes in the weather.

Distribution and Storage Testing

A package may be distributed and stored many times before it arrives to its final consumer. Many of the test procedures focus on corrugated shippers filled with packaged products loaded onto a pallet. There are other situations that require their own unique distribution and storage testing: bulk shipments of empty packages arriving at the manufacturing facility, pallet loads or individual shippers that are modified at the retail store for displayability, or mixed load shipments where a variety of products in various sized shippers are placed together on one pallet.

Distribution tests can be conducted by using real-time actual road ship test methods or by simulating distribution in a lab. It may be best to do preliminary tests in a lab and final approval tests via an actual road distribution test. ASTM has a specified test for simulating distribution in a lab. Many companies develop their own simulated vibration sequence or will alter the ASTM compression or drop testing portions to better model their specific situation. Some labs are even equipped with the ability to test full pallet loads of product at specific temperatures and humidity. Most companies will not have such elaborate facilities, but an independent test lab may. Actual road tests may provide more accurate data, especially if column testing in a lab is the only test option. It is difficult to model shipment of product over mountains or to test various pallet patterns for stability in a lab. To conduct an actual road ship test, a packaging engineer needs to work with the company's distribution network to find a suitable shipping destination. Several locations at various times of the year may be needed depending on the temperature control of the truck and the fragility of the product/package system. To minimize noise in the data, it is best to conduct a ship test using product that is produced and packaged at the manufacturing facility. If the final packaging container is a corrugated shipper, the shippers should be produced on a press. Corrugated shippers fabricated in a lab will not be representative of the final shipper's compression strength.

Simulating storage using compression test equipment in a lab is an option, but the effects of long-term compression can often only be understood with actual pallet load storage testing. The effects of pallet patterns, uneven weight distribution, pallet deck boards, and climate over time can be difficult to simulate on lab equipment.

Package Machineability

How a package will perform on the manufacturing facility's packaging equipment must also be tested. These tests are entirely package, product, and equipment specific. All equipment must be considered, from the conveyors to the palletizer. An analysis of the packaging equipment systems helps to determine the critical areas for a specific package. The purpose for many of the material mechanical properties tests is to indicate if a package will machine well on the packaging equipment. A paperboard carton requires a specific coefficient of friction or it will slide or jam in the machine; a film must have a certain degree of hot tack or the sealing process will decrease line speed; and a hot fillable plastic bottle requires high temperature stability or it cannot successfully receive a closure after being hot filled.

Project timelines increase greatly when new equipment, particularly the packaging production equipment, is purchased due to the financial approval and qualification processes. The project team also grows to include people to manage the installation of new equipment. In the past, many companies would custom-make their packaging production equipment. Most equipment is now purchased with some features customized to meet specific needs. Researching options and choosing an equipment manufacturer must be done with great care. Input from the manufacturing facility should be obtained, because they will be using the equipment on a daily basis.

Lead times to receive production equipment can be quite long, up to a year or more in some cases. Once the equipment is ready, it should be tested at the equipment manufacturer's facility first. It is easier and less expensive to make adjustments and modifications at the equipment manufacturer's facility than at the product manufacturing facility. Sometimes the equipment takes longer to manufacture than is first stated, creating a sense of urgency and a tendency to rush the equipment qualification process. It is very, very important to test the equipment carefully and extensively. Cutting corners to hurry equipment qualification usually adds time to the project in the long run. Once it has been determined that the equipment meets the prior specified criteria, it can be crated and shipped to the manufacturing facility where it is to be further tested.

The need for in-line and off-line tests at the manufacturing facility must also be determined. Such tests may include leak testing, metal detection, seal strength, fill level, label placement, code date legibility, and closure removal torque. If every package must be tested, then an in-line test with an automatic correction or alarm system is ideal. Off-line testing requires labor and could be a concern for the plant.

Manufacturing facilities have tightly controlled cost centers. The total packaging system will need to meet an efficiency rate to be cost effective as line efficiency is directly related to product cost. A labor intensive, slow packaging process may produce a fine package but may also be cost prohibitive. There can be hidden costs as well, such as quality issues due to employee fatigue and a greater potential for workers' compensation issues.

Retail Concerns

Additional tests may need to be done to ensure the product will meet the demands of the retail store environment. Retailers are becoming more specific regarding case count, maximum case weight, and shipper design. If the shipping case is difficult to open, the stock people may cut the shipper open with a knife, potentially damaging the internal packages in the process. The height of store shelves or size of display bins must also be considered when designing a package. The type of lights used in retail stores, especially for products susceptible to light oxidation, may be a concern. Some products can be oriented on a shelf with more than one facing. The layout of the graphics then needs to incorporate two facings. If products are to be stacked on the shelf, they must be stable. Some packages can incorporate stacking lugs to add stability to stacked packages.

Consumer Concerns

When proving the functionality of a package, there are many consumer concerns besides providing a damage-free, fresh product that must be taken into consideration. Ease of opening, handedness, storage, and tamper evidence are a few. Some consumers, especially children and the elderly, have specific concerns.

Crisis Management

Issues often arise, and even emergency situations will occur, during the course of a packaging project. Murphy's Law is extremely applicable to the job of a packaging engineer. One proactive approach is to brainstorm for the unthinkable that can go wrong on a project and conduct appropriate "Murphy testing." Just like the brainstorm ideation session, various people are requested to meet to ideate for possible project glitches. Every environment and every process that the package is exposed to is dissected for any potential concerns. Tests and/or safeguards are put in place to help eliminate the probability of the unthinkable from happening.

Not all trouble situations can be planned for. When a crisis occurs, assess the situation and provide options for dealing with the unforeseen.

(1) Understand and identify any risks associated with the crisis.
(2) Gather useful and pertinent information.
(3) Determine possible solutions and options.
(4) Determine the effect each option will have on the project.
(5) Make recommendations and obtain team consensus.
(6) Take action.

Plant Trials

A plant trial is the process of running a package on manufacturing equipment for a specified period of time to gain package performance information. Plant trials may be very short, just long enough to obtain

Plant Trip Notice: (Project Name)

        Date: (date memo is written)
          To: (list names of everyone at the plant involved in the trial)
      From: (your name)

Plant Location: (city, state or plant name if applicable)
Dates of Trial: (month/day/year, first, second or third shift)
Trial Attendees: (your name and all others accompanying you)

Trial Objective: State briefly what you hope to accomplish and learn from the trial.

Trial Specifics: Provide detailed information on the following:
- Trial Packaging Materials: (List all packaging materials being trialed, control and test. If supplies are being shipped to the plant, provide information on how the material is packed and labeled, when it is expected, how much is being shipped and to whose attention it is addressed. Provide a dock or shipping number if one is available.)
- Equipment Needs: (Explain any special equipment needs and if any equipment must be installed prior to the trial.)
- Trial Duration: (Provide start time and length of time needed to complete trial.)
- Special Concerns: (Provide any special concerns or potential line down time that could result from the trial.)
- Requested Product: (How much product is to be collected; how it is to be labeled; and when, where and how it is to be disposed of after the trial is completed.)
- Charge Number: (Provide for payment of the trial, if required, via a department charge number for PO number.)
- Give your phone number for any question or comments.

.cc (copy all attendees and anyone else applicable such as your manager, team members, suppliers, etc.)

Figure 6.2. Example of a plant trip notice form.

70

packages for lab tests, or very long and extensive, to gather specific data and assurances as to how well the package machines on the packaging equipment. Whatever the purpose for the trial, communication and planning are key to achieving the end objective with fewer headaches.

Plant trials can be very costly, so they should be well planned so as not to be wasteful of materials or time. Plant trials also interrupt the plant's production process and can affect their daily productivity numbers. Due to fast production line speeds, many plant trials will be quick and thus somewhat chaotic. A well-planned trial will be appreciated. A Plant Trip Notice, which is a written request for a plant trial, should be distributed to those involved to ensure everyone has the same information. An example of a trip request is provided in Figure 6.2. A pretrial meeting or telephone conference to review the trial can also ensure the plant trial will go more smoothly. There should be enough people involved in the trial so that the packaging engineer can devote his/her attention to what is being tested rather than be concerned that the test packages are being pulled off the packaging line and are being properly labeled as test product. Showing appreciation to all those who participated with the trial as well as sharing trial results in a written memo helps to keep the plant supportive of a project.

FINAL APPROVAL

The same decision-making group that met for the first project evaluation meets again to determine if a package will proceed to market or not. Marketing once again is responsible for showing the financial viability for a project. The packaging engineer is ultimately responsible for ensuring a package's performance. Full utilization of available resources, team members, manufacturing facilities, suppliers, lab test facilities, literature, and company experts, will ensure that the final approval decision for a package is the correct one. Any risks associated with approving a package must be clearly understood. A good time to conduct a packaging review meeting is a few weeks prior to the final approval meeting.

Again, financial resources are usually required for a package launch, but there is much more at risk than dollars when making a package launch decision. The brand image of the product may be at stake and many people's careers may be adversely impacted if the package launch does not go as well as planned.

PROVING FUNCTIONALITY: CHAPTER SUMMARY

- There are three properties of packaging materials: barrier, optical, and mechanical.
- Finished package tests are determined by the requirements of the product and the environment to which the package is subjected.
- Shelf life testing is done by using the finished package and product and testing them in the intended storage environment over the life of the product.
- Through accelerated shelf life testing and experimentation, it may be determined that exposure to extreme conditions for a shortened time period models the effects of exposure to normal storage conditions over the normal product shelf life.
- A plant trial is the process of running a package on manufacturing equipment for a specified period of time to gain package performance information.
- A plant trip notice is a written document provided to a manufacturing facility requesting a plant trial. The document states the trial objective and all trial specifics such as timing, who will attend, materials to be tested, and what is to be done.

PACKAGING PROJECT EXAMPLES

Scenario A: Credo's Productivity Project

Bill develops the following test plan for proving functionality of the five packaging productivity ideas.

Step 1: Determine the Package Performance Criteria

When conducting the feasibility assessment, Bill determined the package performance criteria. A new packaging film must

- not affect the flavor quality of Credo's
- maintain seal integrity

- machine well on all current plant packaging equipment
- maintain the current graphics quality
- be acceptable to consumers

Step 2: Determine the Test That Must Be Completed for Package Approval

Bill uses the performance criteria to determine the tests that must be completed. He also indicates how each test will be measured. Current and test packaging film will be used for all the tests.

Credo's Productivity Project Test Plan

Test	Measure
Shelf life (a) 45°F stored in shipping case (four months, measured monthly)	Flavor profile and color at parity to current
(b) 45°F under dairy case lights (two weeks, measured daily)	
Sensory Test at two and four months	Flavor, texture, and freshness at parity to current
Seal strength Using serrated Instron jaws, 2 inches/minute	Minimum 600 g/in
COF (kinetic)	0,2 to 0.25
Ink adhesion (a) Sutherland Rub, 300 passes (b) Tape test	Graphics not affected No ink pick or loss
Machineability Extended plant trial (one month)	Line efficiency maintained, knife wear not affected, no increase in product loss allowance
Storage and distribution (a) Simulated storage and distribution	Maximum of 1% damage, 100% inspected

(continued)

Test	Measure
(b) Road distribution	Maximum of 1% damage, 100% inspected
(c) Storage (inspect after one month)	Maximum of 1% damage, 100% inspected
Frozen testing 0°F and 45°F, 20 minutes on vibration table	At parity to current
Code date Visual check at plant trial and inspect product from storage and distribution test	Legibility at parity to current
Consumer test	Test film acceptable to consumers, perception of product quality unaffected

Step 3: Determine What Actions Must Be Taken for Each Test

Bill now needs to determine

- how much test packaging material he needs
- how many packages for each test he will need

Bill looks over his test plan. He has five test film options that he has identified for the project. He realizes it will be very costly to fully test all five of the films. He also surmises that it would be cost prohibitive to conduct the one month extended plant trial unless he can ship the product to consumers. A majority of the tests must be completed before Bill will feel confident in shipping any of the test film to consumers. It would also be time prohibitive to test all five test films for a one month plant trial each. Bill decides to perform all of the tests using each of the five test films except the extended plant trial, the full pallet road distribution test, and the storage test. Once the first round of tests is completed, he will approach his project team with a recommendation of one test film for the remaining tests.

To obtain packages for the first round of tests, Bill will need to travel to the manufacturing facility and conduct a plant trial. He will need to acquire product packaged in each of the five test films and product packaged in the current or control film. Although run time will not be long enough to truly assess machineability, the trial will provide a first indication of how well each film will machine on the packaging equipment.

To conduct the initial trial, Bill will need to obtain samples of the five test films from his film supplier and calculate the number of packages he will need to complete the first round of tests (all tests excluding the extended plant trial, road distribution test, and storage test). He adds in a few extra packages for each test and calculates that 45 cases of 24-count shippers of product will be needed. Three cases will be needed for shelf life testing; three for sensory testing; three more cases for coefficient of friction, ink adhesion, and machineability testing; 16 cases for simulated distribution; 10 for freezer testing; and another 10 cases for the consumer testing.

Bill knows that there are 16,000 impressions on a roll of the current packaging film. He also knows that the film supplier's process dictates they produce a full master roll of film when making test film. Thus, he surmises that one roll of each of the test film variables will be more than sufficient for the 45 cases, or 1,080 impressions, needed.

To calculate the amount of film needed for the one month extended line trial, Bill must first know Credo's manufacturing line speed. He calls Sandy in operations and finds out that the line speed is 180 packages per minute running 16 hours a day on four packaging lines. Bill bases his calculation on the current film having 16,000 impressions/roll and estimates needing 1,296 rolls of film for the one month extended plant trial.

$$180 \text{ pkgs/min} \times 60 \text{ minutes/hour} \times 16 \text{ hours/day}$$

$$= 172,800 \text{ packages/day/line}$$

$$172,800 \div 16,000 = 10.8 \text{ rolls of film/day/line}$$

$$30 \text{ days/month} \times 10.8 \text{ rolls/day/line} \times 4 \text{ lines} = 1296 \text{ rolls of film}$$

Step 4: Determine Timing for Each Test

The first round of tests must be completed prior to the one month extended trial. The longest lead times for the first round of tests are shelf life and sensory which each take four months to complete. The one month extended trial can then be conducted. The distribution and stor-age tests can be initiated at the beginning of the one month trial and will be completed by the end of that month. The missing information is the timing to obtain film for the first trial and then the one month extended trial. Bill phones his film supplier to get their timing for supplying the needed packaging materials. With their information, Bill is able to de-velop a timeline for his test plan (see Table 6.1).

Material and Finished Package Testing

Now that Bill has a test plan, he takes action. He has the five test film options produced. He requests a trip to the plant and schedules the plant trial. At the trial, all five films appear to machine well on the equipment. The line operator commented that one of the variables slipped around a bit more in the forming block, but felt the run time was too short to be sure. The code date ink adhered to all film samples. Bill takes notes of all of the trial details and returns to the office feeling good. He submits product to the sensory and shelf life labs and provides cases of each variable to Marge for the consumer test.

In the next three weeks, he conducts the seal strength, coefficient of friction, ink adhesion, and freezer tests in the lab. The monolayer poly-propylene film out-performed the current film when frozen. One film experienced a lot of cracking from the frozen test and another tested high for coefficient of friction. Two months pass. One of the test films appears to be giving off a fruity flavor to Credo's. To date, the project is on track, and some of the films are testing comparably to the current packaging film, especially the monolayer polypropylene film. The monolayer film will provide the largest productivity savings. Bill is feeling confident.

Crisis Management

Two and a half months into the testing, John from marketing runs into Bill. John mentions that marketing has decided to revise the graphics for Credo's, and he happens to have the latest revisions with him. Bill looks them over and notes that a new silver color has been added to the exist-ing graphics color profile. It is very shiny and metallic looking. The sil-ver was added for a more modern, upscale feeling, explained John. Bill compliments John on the new graphics, but he also tells John he is going

Table 6.1. Credo's Test Plan Timeline.

Activity	Month 1	Month 2	Month 3	Month 4	Month 5	Month 6	Month 7	Month 8	Month 9
Obtain one roll of five test films	xxxx	xx							
Conduct line trial		x							
Conduct first round of tests		x	xxxx	xxxx	xxxx	xx			
Choose one film						x			
Obtain one month supply of selected film							xxxx		
Conduct one month extended trial and finish remaining tests								xx	xx
Final team approval									xx

to call his film supplier to discuss the new color, just to be on the safe side.

Bill discovers that the ink used to create the silver varies slightly in chemical composition from the other inks used in the Credo's graphics. The substance that makes the silver so shiny is approved only for indirect use in food packaging materials. The film supplier comments that it is unlikely the new ink will affect the product. Bill feels that may be true as long as the ink is sandwiched between two layers of polypropylene, but he is concerned about its use in a reverse printed monolayer film. He is six weeks from completing the first round of tests and choosing a film for the one month extended trial. The new silver ink could jeopardize the project and, at the very least, delay it.

Bill follows the crisis management steps:

(1) Understand and identify any risks associated with the crisis.
   • The monolayer film that is being tested has a layer of heat seal adhesive between the ink and Credo's. Does the heat seal act as a functional barrier between the silver ink and the product?
   • Will the silver ink impart any off-flavor to Credo's in multilayer or monolayer film structures?
   • Is silver ink adhesion comparable to the current ink or could it possibly flake off?

(2) Gather useful and pertinent information.
   Bill speaks to the film supplier and discovers that the heat seal adhesive is not considered a functional barrier for the silver ink. Bill asks if there is another silver ink option that could be used, but there is not. Next, Bill phones John and explains the situation. He asks John how crucial silver ink is to the new graphics. Marketing is insisting on keeping the silver ink for a tie-in to promotions that are already moving forward.

(3) Determine possible solutions and options.
   • Eliminate the silver ink.
   • Eliminate monolayer as a productivity savings option.
   • Determine a method to achieve a functional barrier between the ink and the product in a monolayer film without affecting the productivity savings.

(4) Determine the effects each option will have on the project.
   • Eliminating the silver ink would involve pulling promotional items that are already with the sales force, and creating new promotions at the last minute.

- The largest productivity savings can be realized with the monolayer film. Without it, the Credo's project will under deliver on savings and a price increase will likely result causing potential customer loss.
- Bill phones his film supplier and inquires if there is a method for achieving a functional barrier between the ink and the product. The supplier has been investigating an idea of applying a lacquer coating between the ink and the heat seal layer. He just obtained a cost estimate for the idea and was about to call Bill. Bill learns that the lacquer coating will solve the silver ink direct contact issue and will only marginally affect cost.

(5) Make recommendations and obtain team consensus.

Bill recommends obtaining a roll of monolayer film with a lacquer coating and silver ink in the graphics and submitting it to the test program. If all goes well, the film will allow for the completion of the tests four months behind the project's scheduled completion date. This "new" monolayer film, now labeled option F, still has the largest productivity savings potential. He also recommends testing a downgauged multilayer film with silver ink, option G, as a contingency, just in case the monolayer option fails.

Although the other productivity film options do not have a direct contact concern regarding the silver ink, there is still a chance that the silver ink could cause an off flavor. Bill feels the risk is low, but he can't be sure. He recommends conducting accelerated shelf life testing and comparing films with and without the silver ink. These tests are not foolproof but could give a better indication of the risk of proceeding with the silver ink without full shelf life testing. The most conservative approach would be to obtain all the test films with the new silver ink and test them again.

The team agrees that the new monolayer film, option F, must be tested. They agree to meet again when the accelerated shelf life test results are available. Results from the first round of tests will also be available at that time. A decision will be then be made either to proceed with one of the test films without completing full shelf life testing with the silver ink or wait another four months.

(6) Take action.

Bill requests that the film supplier produce a roll of the monolayer film, option F, and the downgauged multilayer film, option G. As soon as they are available, he will request a plant trip and have packages made on the manufacturing equipment. He will then proceed with testing the films according to the test plan and conducting accelerated shelf life tests.

Final Approval

Bill completed all of the tests on the original five film options. He obtained Credo's packaged in the films containing silver ink in the graphics and completed some of the tests using those packages. The team reassembled to discuss the results to date and to decide if they should proceed without full shelf life testing with regard to the silver ink. Bill summarized the data in Table 6.2.

Bill walks the team through all the tests that have been done and again explains the current situation regarding the silver ink concern. Although accelerated shelf life data indicates no difference between film printed with the silver ink and film without it, Bill prefers to take the conservative approach and complete full shelf life tests. He explains that the project is currently a month behind the schedule they had established when the project began eight months ago. One of the criteria for success was to complete the project within an 18-month time period. If they delay the timeline another three months to complete shelf life testing on the films with silver ink, they can still finish the project within the 18-month time period. The team agrees with Bill's recommendation. They will meet one more time when shelf life testing is completed.

Bill is able to deliver good news to the team three months later. The monolayer film, option F, successfully completed shelf life and sensory testing. The team agrees that the monolayer option F is to be tested for one month in an extended plant trial. Pending a successful extended trial, Bill will notify purchasing that a new film has been qualified. Purchasing will then convert to the monolayer film on a flow through basis, meaning that the current film inventory will be fully utilized first. The conversion process is expected to take two months.

Scenario B: Oula Package Redesign

Susan obtains the necessary funds once the project is approved to initiate the production tool for "Stacking Package A" and to purchase the

Table 6.2. *Credo's Test Results Summary.*

Film Option	Film Gauge	Est. Annual Savings	Performance	Recommendation
A. Monolayer pp	1.15 mil	$2.0 MM	At parity to current	Pursue only if silver ink is omitted from graphics
B. Alternate material, monolayer	1.10 mil	$1.9 MM	Off flavor detected in shelf life	Discontinue
C. Decreased pp in multilayer film, alternate pp supplier	1.15 mil	$1.6 MM	At parity to current	Pursue only if silver ink is omitted from graphics
D. Decreased pp in multilayer film	1.15 mil	$1.4 MM	At parity to current	Discontinue, greater savings potential with other options
E. Alternate material, multilayer film	1.20 mil	$1.4 MM	Potential package forming issues	Discontinue
F. Monolayer pp with lacquer and silver ink	1.15 mil	$1.95 MM	Accelerated shelf life indicates no issues, all other tests at parity to current	Complete four month shelf life testing
G. Decreased pp in multilayer film, alternate pp supplier with silver ink	1.15 mil	$1.55 MM	Accelerated shelf life indicates no issues, all other tests at parity to current	Complete four month shelf life, pursue as a contingency to option F

needed change parts for the packaging equipment. The cavities on the production tool will not be cut until Susan can complete additional tests, but a portion of the tool can be started. Susan also develops a test plan for qualifying the package.

Step 1: Determine the Package Performance Criteria

The Stacking Package must:

- protect the product from moisture loss
- be chemically compatible with the product
- perform at parity to current in simulated and actual distribution tests
- machine well on all packaging equipment
- maintain closure integrity
- comply with all specified package dimensions and tolerances
- consistently match the specified resin colors

Step 2: Determine the Tests That Must Be Completed for Package Approval

Susan now uses the performance criteria to develop the test plan. The resin that is used to produce the current package will also be used to produce the Stacking Package. There have been no chemical compatibility issues in the past, and the resin has excellent moisture barrier properties. Had Susan selected a different resin, she would have completed the chemical resistance tests in the feasibility portion of the project. She plans to conduct the test during the proving functionality phase more as a cursory step, since there is a usage history with the chosen resin.

Oula Stacking Package Test Plan

Test	Measure
Mocon package moisture barrier transmission	At parity or better than current
Chemical resistance test	No stress fractures after one week exposure

Test	Measure
Storage and distribution	
Preproduction: (empty packages)	Maximum of 0.05% damage, 100% inspection of empty packages in reshippers
(a) Actual distribution	Inspection of empty packages in reshippers
Postproduction:	
(b) Simulated distribution	At parity or better than current, 100% inspected for damage and closure integrity
(c) Actual distribution	At parity or better than current, 100% inspected for damage and closure integrity
(d) Vibration	At parity or better than current, 100% inspected for damage and closure integrity
Machineability	
(a) Plant trial using packages produced on unit tool	No loading, filling, capping, conveying or shipping issues
(b) Plant trial using packages produced on production tool	Minimum 90% line efficiency
Closure integrity	Removal torque = 14 in·lb ±2
Label adhesion	100% adherence, no peel
Label abrasion	
(a) Sutherland Rub, 150 and 300 passes	No visual abrasion
(b) Accumulator table test 1 hour	No visual abrasion
Dimensional integrity Three packages from each cavity (16 total cavities):	
(a) Optical comparator to measure dimensions	See blueprint for specified dimensions and tolerances

(continued)

Test	Measure
(b) Micrometer to measure wall thickness (every 1/4 inch horizontally and vertically)	See blueprint for specified wall thickness and tolerances
(c) Overflow capacity	96 cc each container
Maintain resin color standards Resin colors must match color chip	To be checked at package manufacturer at beginning of each batch

Step 3: Determine What Actions Must Be Taken for Each Test

Susan has enough package samples to initiate the Mocon tests right away. She will need to obtain "dog bone" Instron samples from the resin supplier to conduct the chemical resistance tests, and she will need packages produced on the unit tool for the first plant trial. Susan wants enough packages to run for two hours. Based on line speed, Susan will request 12,000 packages from the supplier. If the trial and first round of tests go well, Susan will give the supplier approval to complete the production tool. Upon completion of the production tool, she will have packages from each cavity sent to her for dimensional integrity testing. Once approved, Susan will request that 48,000 packages be produced for another plant trial.

Step 4: Determine Timing for Each Test

Susan establishes a timeline for completing the test plan (see Table 6.3). She knows she may need to add additional line trials if issues occur, but feels they can be conducted during the time the production tool is being produced, thus not affecting the stated package approval date.

Line Trials and Murphy Testing

After several months and long hours spent working to correct conveying, cartoning, and capping equipment issues on one of Oula's three packaging lines, the Oula Stacking Package is finally qualified. As line speed increased during the final line trial, a potential problem was dis-

Table 6.3. *Oula Test Plan Timeline.*

Activity				Timing				
	July	Aug.	Sept.	Oct.	Nov.	Dec.	Jan.	Feb.
Material tests	xxxx							
Initiate prod. tool	xxxx	xxxx						
Obtain change parts	xxxx	xxxx						
Install change parts			xxx					
Initial line trial			x					
Package tests			x	xxxx				
Finish prod. tool					xxxx	xxxx		
Prod. tool approval (dimensional tests)							x	
Final line trial							x	
Package tests							xx	xx
Final package approval								xx

covered. Susan monitored closure torque removal values throughout the line trial. Once a faster line speed was achieved, she noted that the caps were too loose. She checked closures applied on containers from each of the capping heads and found them all out of specification. While a line technician made adjustments, Susan wondered if the increased line speed could have affected the capping system. The line started back up and the removal torque readings were within specification; they then became too loose and remained so.

The line was shut down once again for more adjustments. Again, the removal torque readings were within specification for a short time when the line started back up and then went out. Susan and the line technician were discussing the situation while looking up at the closure hopper when a puff of steam rose out of the hopper. The line technician explained that the caps are exposed to steam in the hopper prior to capping to warm up the cap liners so they are pliable. Susan guessed that the steam exposure time had decreased as the line speed had increased and was causing the loose cap situation. The line technician agreed that it may be a possibility. He adjusted the hopper to hold a greater number of closures and opened the steam valve. The line started up, and Susan checked closure removal torque for the next half hour, finding that the readings remained within specification. Susan suggested to the line supervisor that a steam exposure requirement for closures in the hopper be established. That afternoon, she specified the optimum time and steam valve setting and added it to the packaging line guidelines.

Susan was concerned that there might be other factors that she was unaware of, like steam exposure, that could cause problems. She formed a "Murphy search" team consisting of the line supervisor, line technician, and a crew worker and asked that they discuss all steps of the packaging line with her. During their discussion, they listed potential concerns and noted any actions that should be taken. They found an area on the conveying line just before the labeling station that was periodically misted with mineral oil. It was unlikely that the oil's spray nozzle would ever get bent, but if it did, it would spray on the packages. Susan added an oil test to her test plan to determine if a fine mist of oil would inhibit label adhesion. There was also a remote chance that packages could remain on the accumulator table for a longer period of time than the current package did. Susan decided to lengthen the accumulator label abrasion test she had originally planned. Susan learned a great deal about the new packaging line through the Murphy team experience and now feels more confident about the project. She also requests that her packaging supplier brainstorm for potential Murphy situations. Susan completes a project review prior to the final project approval meeting.

Final Approval

The team meets again with management. A video of the Stacking Package on the packaging line is shown while Susan explains the steps that had to be taken to get line efficiency to 90%. She also explains that line efficiency is expected to improve once the manufacturing facility is producing the new package on a full-time basis. Package test data are presented along with a review of the consumer test results. Susan also discusses the Murphy team she formed and their findings. All of the criteria for success factors have successfully been met. The team's recommendation that the new package be launched as soon as possible is supported by management.

Scenario C: Carton Crisis

To qualify a new carton coating supplier, Kathy develops a test plan.

Step 1: Determine the Package Performance Criteria

- cartons must feed one at a time into the setup and gluing mechanism and run efficiently throughout the packaging system
- carton coating cannot adversely affect graphics
- carton seal strength must be maintained

Step 2: Determine the Tests That Must Be Completed for Package Approval

Kathy creates a test plan and uses the current carton as a basis for measure.

Carton Test Plan:

Test	Measure
To be conducted prior to plant trial:	
• COF	current carton = 0.35 kinetic

(continued)

Carton Test Plan:

Test	Measure
• Sutherland Rub	at parity to current, visual comparison
• glue adhesion	at parity to current
• gloss	at parity to current

To be conducted during plant trial or using packages from plant trial:

• code date	legibility at parity to current (to be visually checked during plant trial)
• machineability	cartons feed one at a time into set-up and gluing, ensure no other run issues
• simulated lab distribution	at parity or better than current (inspect for product damage and carton abrasion)
• seal strength	at parity or better than current

Step 3: Determine What Action(s) Must Be Taken for Each Test

Kathy looks over her test plan and calculates the number of packages she will need to complete the tests. She determines that she will need 30 cartons of each variable to complete the tests prior to the plant trial. She plans to test the current carton versus cartons with the new coating. She also has the defective cartons that the manufacturing facility sent to her which she plans to test as a worst-case scenario.

Kathy will retain 15 cases of product produced from the plant trial to conduct the lab tests. She now needs to determine how many cartons will be needed to conduct the plant trial. The equipment speed is 45 cartons a minute, and Kathy would like to have enough cartons for a two-hour trial. She determines that she will need 5,400 cartons. For a buffer, she decides to request 6,000 cartons.

Step 4: Determine Timing for Each Test

Kathy feels the test plan can be completed by the dates she indicated in the September 4 summary page she presented to her manager.

Kathy also needs to determine the right coefficient of friction target and acceptable range for the carton. She phones her carton supplier to discuss the number of cartons she will need for tests, the project timing, and a method for determining carton coefficient of friction requirements.

Material and Finished Packaging Tests

The carton supplier is able to produce cartons with an alternate coating that they believe will function at parity to the current carton coating. They supply carton samples to Kathy, and she completes coefficient of friction, Sutherland Rub, and glue adhesion material tests. Test results indicate that the alternate coating performs similar to the current coating. Kathy gives the approval for cartons to be shipped to the manufacturing facility for a plant trial. The carton supplier also provided Kathy with 200 carton samples with varying coefficients of friction: 0.19, 0.25, 0.32, 0.35, 0.41, and 0.53. The supplier was able to control the coefficient of friction through different coating processes and with additives. Kathy schedules a plant trial planning to test the alternate coating for a two-hour trial period. She will also run the cartons with varying coefficients of friction.

Kathy arrives at the manufacturing facility for the carton trial. Her carton supplier is also in attendance. She surveys the packaging line as the workers are setting up for the trial. She picks up a few of the cartons that are being loaded into the machine. They feel rough to the touch. Kathy is concerned that something is different with the cartons that have been shipped to the plant versus the ones she tested in the lab. She asks the man loading the cartons if they are indeed the test cartons for the trial. He indicates that they are the cartons that were brought out of storage. He assumes they are the ones he is supposed to load for the trial. Kathy inspects the pallet of cartons and notices there is a "hold for inspection" tag on the pallet. Under that tag is a lot number, KL812. She finds the line supervisor and explains that the wrong cartons have been brought out for the trial. The supervisor checks into the situation and finds that the forklift driver misunderstood which cartons the line operator had asked him to deliver.

After a little searching, the test cartons were located and brought to the packaging line. The line was reloaded with the correct cartons. It would only be a few minutes into the trial before they would know how well the alternate carton coating would perform. The trial begins and the cartons machine without any issues. Everyone agrees that the alternate carton seems to run well. Kathy watches the equipment for about 15 minutes to ensure there are no issues.

The packaging line has three separate mechanisms that setup, glue, and load the cartons. Kathy decides to let two of the mechanisms continue to run the cartons with the alternate coating, and on the third, she will test the cartons with varying coefficients of friction. She is able to tell very quickly which cartons can run on the machine and which ones slide too much or jam causing downtime. Kathy collects all of the data she needs and successfully completes the trial. She phones her manager with the good news.

Final Approval

Kathy records the findings from the trial in a memo and distributes it along with a recommendation to approve the alternate coating. She indicates that her next steps will be to modify the carton packaging specification to include a second approved coating. She will also indicate on the specification a coefficient of friction requirement. From the tests conducted during the trial, she has learned that coefficient of friction can range from 0.30 to 0.41 without any equipment issues.

CHAPTER 7

Package Launch

PRODUCTION START-UP

THE PRODUCTION START-UP of a new package is a very exciting and anxious time. Up to this point, the packaging engineer has been central to the project's forward movement. Although the packaging engineer is still actively involved, several of the other team members will now take on a larger role as the project moves into a conversion stage to launch the new package.

The purchasing and the asset management groups, in particular, become integral to the package launch. They will coordinate the use of old packaging materials and ensure that the correct amount of new materials is in supply and on time for the package conversion. Purchasing is responsible for ordering the new supplies and managing the existing inventory of old packaging materials, either utilizing them or writing them off the accounting books as a loss. There are usually many components to a package, and the dispensation of all the materials must be carefully coordinated. The purchasing group must be kept up to date on a new package's targeted production launch date so that launch delays do not occur due to having large inventories of old packaging materials on hand.

The asset management group gauges expected product sales. Prior sales records of existing products can be used to estimate production requirements and cycles, and in turn, the needed packaging materials. If the package is for a new product, sales volumes must be predicted.

Once the final approval has been given for the project, there may be additional packaging manufacturing lines to install and other preparations to be done that require heightened involvement from the manufacturing facility. The packaging engineer's involvement is still great, but essentially, the project is being handed over to the manufacturing facility. The burden is now theirs to manufacture the new package on a daily basis and to do so efficiently. Technical support staff from equipment and material suppliers should attend production start-ups just in case issues arise.

Some people are averse to change. Project commitment can be enhanced if people feel involved and important. A kick-off meeting held at the manufacturing facility prior to a package launch provides an opportunity to motivate and share information. Testing requirements, critical package concerns, specific processes, and goals for the day, such as line speed or product loss numbers, can be reviewed and discussed.

Marketing and manufacturing factors dictate the process of introducing a new package to consumers. Some packaging conversions are done a little at a time, called regional roll-outs. This is done by converting the package in one market region, and then a short time later, converting another region, then converting another until all markets are receiving the new packaging. Marketing may want to do a regional roll-out if they feel there is some risk involved with consumer acceptance of the package. They can monitor packaging performance and make necessary changes to advertising and promotional programs, or discontinue the package launch if needed. A regional roll-out may be necessary if packaging equipment changes are required at several manufacturing locations, and the equipment is converted one location at a time. An inventory prebuild may be needed if the manufacturing facility must shut down to convert equipment for the packaging change or to compensate for low production rates during the new package start-up phase. In other instances package conversion is done all at once for all market regions.

MONITORING PERFORMANCE

How well a new package performs either technically or with consumers will determine if the project truly meets the original goal as stated in the team charter. There are several ways to measure a package's performance. The nature of the packaging change will direct which method is most suitable.

- *consumer response:* Many products provide a toll free number or address on the package so consumers can easily provide feedback, both positive and negative. For every one consumer who lodges a concern or complaint, there are many others who will not take the time to do so. Thus, the actual number of happy or dissatisfied consumers is greater than the number of responses received. The number of consumers who respond tends to increase whenever a packaging change is made.

- *sales volume:* Analysis of product sales trends during the months after a new package is launched can indicate the success of a packaging change. Some packaging changes may be invisible to consumers and thus will not be reflected in sales volume data.
- *test data:* Testing the package after it has launched can verify that the change is meeting the stated goal. Packages may be tested at the manufacturing facility, during the distribution cycle, or even off the store shelf to measure how well a new package is performing.
- *manufacturing records:* Manufacturing facilities keep daily records of production rates, production downtime, packaging loss allowances, packaging usage, packaging quality issues, and equipment maintenance that can be indicators of how well a package is performing.

PACKAGE LAUNCH: CHAPTER SUMMARY

- A kickoff meeting is held at the manufacturing facility prior to a package launch providing an opportunity to motivate and share information. Testing requirements, critical package concerns, specific processes, and goals for the day, such as line speed or product loss numbers, can be reviewed and discussed.
- A regional roll-out is done when a new package is introduced a little at a time by converting one market region and then converting another region a short time later, then another, until all markets are receiving the new package.
- An inventory pre-build is when excess product is produced to cover future production needs while the manufacturing facility shuts down to convert equipment for a packaging change, or to compensate for low production rates during the new package start-up phase.
- Methods used to monitor packaging performance after launch are

Consumer Response, Sales Volume, Test Data, and Manufacturing Records.

PACKAGING PROJECT EXAMPLES

Scenario A: Credo's Productivity Project

After one month of using the monolayer film, option F, the manufacturing facility provides their approval for the film. Line speed, product loss allowances, and all maintenance procedures were maintained during the extended trial. Bill has completed all of the necessary tests, and the team agrees to convert all packaging to the monolayer film. The conversion process is a fairly simple one from the packaging engineer's perspective. Once inventories of the old film are depleted, they will be replaced with the new monolayer film. The film conversion process takes two months to complete. The project is completed after 14 months of effort. Productivity savings are $1.95 MM annually.

Bill will monitor the monolayer film using consumer response data. He will focus on packaging comments, watching for any changes that could be due to the film change. Bill will also contact the manufacturing facility several times in the next few months for input. The manufacturing facility will be monitoring the film's performance by tracking product loss rates, film quality defect rates, and equipment knife wear.

Scenario B: Oula Package Redesign

Susan has successfully completed all packaging tests and line trials to qualify the new Oula package. Transition to the new package will begin immediately. It will take approximately two months to obtain all of the necessary packaging materials and convert the other two packaging lines. Marketing has decided to write any remaining current packaging materials off as a loss in order to bring the new package to market faster.

Susan plans to attend the start-up of the new package and clears her schedule so that she can spend a week at the manufacturing facility. Although the package has been qualified and ran well at the line trials, faster line speed is needed to achieve the desired line efficiencies. An Oula start-up team consisting of several manufacturing personnel, the packaging supplier, and Susan is formed to ensure a successful package launch.

The first day of the start-up is a bit chaotic but goes rather well considering the newness of all of the packaging components. Susan started the day by providing doughnuts and coffee at a kickoff meeting for the start-up team and the line crew. The line technicians spent much of the day adjusting the packaging line systems to enhance product flow. Susan tested closure removal torque periodically throughout the day and talked with the people on the packaging line to gather their input on possible improvements. After a few days, the line crew is more familiar with the new package and line speed begins to increase. It then becomes apparent that the three lines cannot be manually loaded fast enough to feed the filling equipment. The intention is to automate the package loading system in the following year. The space where the empty packages are manually loaded onto the packaging line is very limited, allotting for only two people.

The start-up team watched the loading operation and then met in the break room to brainstorm for solutions. Automating the loading system will take 10 to 12 months. A short-term solution to improve line speed must be found or the project will not meet its financial hurdles. Everyone on the team had noted that the packages were not orienting properly when loaded. Package orientation had not been an issue at any of the previous line trials and was determined not to be an issue when a "Murphy" review of the supplier's process had been conducted. The two people loading the packages onto the conveying system must now reposition the packages and thus cannot supply the downstream equipment. After some analysis, the team realizes that with some slight alterations to the conveyors, enough space can be made for an additional person to be brought onto the line to orient any packages that are out of position. The packaging supplier has also phoned his manufacturing contact to determine why the packages are not arriving properly oriented. Meanwhile, the addition of another person in the package loading area will keep the line running smoothly.

After several weeks, the line speed has increased and is expected to be at the required rate within another week or two as the manufacturing facility completes all the adjustments and climbs the learning curve for the new system. Susan has also resolved the package orientation issue with the packaging supplier. The supplier took full responsibility for the situation. The orientation issue was caused when an untrained employee replaced the regular employee who had gone on vacation. The supplier explained that the replacement person had programmed the packing controls according to requirements for a different but similar product. They guaranteed it would not happen again. Susan's stay at the plant lasted three weeks instead of one, but the launch of the new package was successful. Package performance will be monitored using all monitoring methods: consumer response, sales volume, test data, and manufacturing records.

Although the new package has successfully been launched, there is still more project work to be done. Susan will now manage the next step of auto-loading bulk packaged, empty containers onto the packaging line. This portion of the project is anticipated to take 12 months to complete and will eventually pay for itself in labor savings.

Six months after the new package launch, consumer response is very positive toward the new package. As was expected, some consumers complained regarding the price increase, but sales have increased more than was anticipated. In the last few months, it has become apparent that some consumers do not understand the container replacement process. Some complaints are still being recorded in which consumers say they cannot finish the two creams at the same time. The team met to discuss the situation and has decided to clarify the verbiage used on the package that explains the replacement process and to place more emphasis on the process in advertising. If consumers continue to complain, the team will decide if further action is needed, possibly reducing the size of one container. Such a project would require Susan's involvement, as a new production tool would be needed for one container, and applicable qualification testing would be needed.

Scenario C: Carton Crisis

Kathy updates the carton packaging specification to include the alternate coating and the coefficient of friction requirement. Purchasing is now able to work with the carton supplier to ensure there are no carton inventory issues. Kathy plans to monitor the consumer response data for any indication that the alternate coating causes issues, although she does not anticipate any.

Grossman, S. (1984). "Releasing Problem Solving Energies." *Training and Development Journal,* pp. 38, 94–98.

VanGundy, A. G. (1985). *The Product Improvement Checklist (PICL).* Norman, OK: VanGundy and Associates, Inc.

Warfield, J. N., Geschka, H. and Hamilton, R. (1975). *Methods of Idea Management.* Columbus, OH: The Academy for Contemporary Problems.

Milton Keynes UK
Ingram Content Group UK Ltd.
UKHW020032071024
449327UK00032B/3039